T0205565

Blockchain Technologies

Series Editors

Dhananjay Singh, Department of Electronics Engineering, Hankuk University of Foreign Studies, Yongin-si, Korea (Republic of)
Jong-Hoon Kim, Kent State University, Kent, OH, USA
Madhusudan Singh, Endicott College of International Studies, Woosong University, Daejeon, Korea (Republic of)

This book series aims to provide details of blockchain implementation in technology and interdisciplinary fields such as Medical Science, Applied Mathematics, Environmental Science, Business Management, and Computer Science. It covers an in-depth knowledge of blockchain technology for advance and emerging future technologies. It focuses on the Magnitude: scope, scale & frequency, Risk: security, reliability trust, and accuracy, Time: latency & timelines, utilization and implementation details of blockchain technologies. While Bitcoin and cryptocurrency might have been the first widely known uses of blockchain technology, but today, it has far many applications. In fact, blockchain is revolutionizing almost every industry. Blockchain has emerged as a disruptive technology, which has not only laid the foundation for all crypto-currencies, but also provides beneficial solutions in other fields of technologies. The features of blockchain technology include decentralized and distributed secure ledgers, recording transactions across a peer-to-peer network, creating the potential to remove unintended errors by providing transparency as well as accountability. This could affect not only the finance technology (crypto-currencies) sector, but also other fields such as:

Crypto-economics Blockchain
Enterprise Blockchain
Blockchain Travel Industry
Embedded Privacy Blockchain
Blockchain Industry 4.0
Blockchain Smart Cities,
Blockchain Future technologies,
Blockchain Fake news Detection,
Blockchain Technology and It's Future Applications
Implications of Blockchain technology
Blockchain Privacy
Blockchain Mining and Use cases
Blockchain Network Applications
Blockchain Smart Contract
Blockchain Architecture
Blockchain Business Models
Blockchain Consensus
Bitcoin and Crypto currencies, and related fields

The initiatives in which the technology is used to distribute and trace the communication start point, provide and manage privacy, and create trustworthy environment, are just a few examples of the utility of blockchain technology, which also highlight the risks, such as privacy protection. Opinion on the utility of blockchain technology has a mixed conception. Some are enthusiastic; others believe that it is merely hyped. Blockchain has also entered the sphere of humanitarian and development aids e.g. supply chain management, digital identity, smart contracts and many more. This book series provides clear concepts and applications of Blockchain technology and invites experts from research centers, academia, industry and government to contribute to it.

If you are interested in contributing to this series, please contact msingh@endicott.ac.kr OR loyola.dsilva@springer.com

More information about this series at http://www.springer.com/series/16276

Dhananjay Singh · Navin Singh Rajput
Editors

Blockchain Technology for Smart Cities

 Springer

Editors
Dhananjay Singh
ReSENSE Lab, Department
of Electronics Engineering
Hankuk University
of Foreign Studies (HUFS)
Yongin-si, Korea (Republic of)

Navin Singh Rajput
Department of Electronics Engineering
Indian Institute of Technology (IIT) BHU
Varanasi, India

ISSN 2661-8338   ISSN 2661-8346   (electronic)
Blockchain Technologies
ISBN 978-981-15-2207-9   ISBN 978-981-15-2205-5   (eBook)
https://doi.org/10.1007/978-981-15-2205-5

This Springer imprint is published by the registered company Springer Nature Singapore Pte Ltd.
The registered company address is: 152 Beach Road, #21-01/04 Gateway East, Singapore 189721, Singapore

Contents

**Introducing Blockchain for Smart City Technologies
and Applications** 1
Rayan M. Nouh and Dhananjay Singh

Chain of Antichains: An Efficient and Secure Distributed Ledger 19
Jinwook Lee and Paul Moon Sub Choi

Blockchain for Intelligent Gas Monitoring in Smart City Scenario 59
Ashutosh Mishra, Rakesh Shrestha, Shiho Kim and Navin Singh Rajput

**Commodity Ecology: From Smart Cities to Smart Regions
Via a Blockchain-Based Virtual Community Platform
for Ecological Design in Choosing All Materials and Wastes** 77
Mark D. Whitaker and Pravin Pawar

**Toward Multiple Layered Blockchain Structure for Tracking
of Private Contents and Right to Be Forgotten** 99
Min-gyu Han and Dae-Ki Kang

**Smart City Transportation Technologies: Automatic No-Helmet
Penalizing System** 115
Ashutosh Agrahari and Dhananjay Singh

**An Overview of Smart City: Observation, Technologies, Challenges
and Blockchain Applications** 133
Vijay Kumar Chaurasia, Alhasha Yunus and Madhusudan Singh

**An Architecture for e-Health Recommender Systems Based
on Similarity of Patients' Symptoms** 155
Valerio Frittelli and Mario José Diván

Introducing Blockchain for Smart City Technologies and Applications

Rayan M. Nouh and Dhananjay Singh

Abstract A smart city is described as far from being an unambiguous concept, changing with each iteration of a much smarter version of residents living or economic activity. However, in recent years the governments around the world have been trying hard to provide the most advanced form of life to residents that their resources can afford. This means that there is no pre-defined definition and any city that is deemed to provide value to its citizens as well as generate benefit for the global society shall be considered smart. The motive of this paper is to provide a comprehensive review on Blockchain technology based smart city and enterprises their motivation, challenges and applications, in order to overcome their services, and production costs. This chapter will further explain about use cases of Blockchain technology for smart city application.

Keywords Smart cities · Blockchain · Technologies · Application and services

1 Introduction

Blobba et al. [1] have explored that a smart city is dependent on the contribution toward global betterment by using the natural resources most efficiently. The city may play host to people belonging to several nationalities who come to it because the city is considered an economic center. The city is also considered smart if it is technologically advanced and is considered the home of every kind of new technology. The point is, the smart city is considered efficient because everything is well synchronized. From citizens to corporations, legal requirements to political forces. The system is time-tuned to produce output by using a small number of resources. The Blockchain can bring betterment because data regarding the public shall be available for view to the public directly. Although. The system's development requires

R. M. Nouh
Department of Computer Science, Dankook University, Yongin, Gyeonggi-do, South Korea

D. Singh (✉)
Department of Electronics Engineering, Hankuk University of Foreign Studies (HUFS), Global Campus, Yongin, South Korea
e-mail: dsingh@hufs.ac.kr

© Springer Nature Singapore Pte Ltd. 2020
D. Singh and N. S. Rajput (eds.), *Blockchain Technology for Smart Cities*,
Blockchain Technologies, https://doi.org/10.1007/978-981-15-2205-5_1

cost but that is one time, the benefits on the other hand in the form of transparency, authenticity and reduced risk of loss of vital data bring innumerable benefits. Smart city's operations depend on data about public demands collected through sensors and evaluating those using models. View ability of the conclusions on the analysis of the data drawn is possible using Blockchain. This is the reason that fore smart cities including San Diego, San Francisco have been outstripped by Tokyo and London. *As smart city hub*, IBM initially described the concept of the smart city through an explanation of three I-based principles. These I's represented Instrumented, Interconnected and Intelligent. An instrument is an appliance, a sensor or any electrical data-dependent system that helps the user by performing a certain function in a more effective and efficient manner. Interconnected means that the entire form of activity relies on data that is accessible by relevant personnel. While as part of the fourth industrial revolution, data drive intelligent system and becomes more efficient and realistic by use of machine learning modelling and analysis performed by AI dependent techniques. Considering these three pillars of smartness, of a living style, it can be said that each of the advanced cities of this world has changed drastically and this change has been contributed by these three pillars. This means that the cities support to the global economy in any form over history have made them rule the world [1].

Albino et al. [2] describes well that best mix between value and resources used to create value is considered smart. These changes the characteristics of a smart city, characteristics are elements that help the city in becoming an epitome in a certain field. The logic makes sense because Singapore is not as technically advanced as Tokyo yet; it is considered the well-planned city of this world. It cannot be said about Tokyo where traffic problems are a mess. Similarly, Germany is economically advanced, while Japan is technologically fast and supports this world through ingenious iterations of smart technologies. This means that tackling any of the external forces (technology, economic, social, ecological, political and legal) in a manner that is better from the current best manager. Makes the city a smart residence or economic activity performer. However the authors have found that the major challenge in creating the smart city amid increasing concerns for the ecological imprint of human activity. Is the merger of technology and resources to reduce the human activity effect on the environment? Sustenance of environment can fund the further expansion of the global economy in the long-term. Bitcoins was presented in the fall of the year 2008 while implemented in early 2009. The technology used for the currency transfers only have fascinated the users since then to use the technique on making available important public data through transparent means. Economic indicators, the population of the country, usage of tax money and other such data is public property hence, its visibility through the decentralized system is forcing the researchers to re-invent Bitcoins. The efforts led toward the development of the Blockchain data system. Data is sharable among the nodes (computers connected to a network) while at the same time it is accessible to all the network users. Users can add data into already stored data however; the user cannot change it because of the chained link between the entered data and already stored data. Further, transparency of the system makes an entire network of users' owner of the data who can accept or reject this change. Additionally, since the data is publicly available and almost immutable this makes the system less risky to the attacks from hackers. The technology also reduces the risk of loss of important personal data since it uses an algorithm that distorts (codes) the details of party and transaction itself [2].

Crosby et al. [3] have observed that Bitcoins success has raised voices for the use of the Blockchain method in the corporate world. The financial world is arcane because it depends mostly on the logics made using difficult financial models and techniques. These models are non-understandable for the professionals and have contributed to the loss of the public's money in the significant financial crises. Crises that first shook this world in 2008 and then again in 2012. Even the multi-national companies including Nissan, Google, VW and Toshiba have cheated the authorities, once again causing massive loss to the public. Forget about the private sector, the public sector that has been endowed with the public's trust to care for them is not immune to corruption either. This menace is due to lack of transparency or accountability, however, concerns regarding the loss of sensitive data have always played afoul of the public. The monstrosity eating at public's fund in the form of corruption or huge corporate frauds could be contained significantly by the use of Blockchain system. Bitcoins technique's implementation has its uses in financial and non-financial sectors, the sectors that conventionally have depended on the third party for enforcement or safeguard. These contracts can now be triggered by the simple occurrence of a pre-set condition. In particular, an application was introduced in 1994 by Nick Szabo the idea was to automatically execute the contracts between the two parties. Blockchain technology in combination with this idea may do wonders and all the contracts that depend on lawyers or bank's escrow services may be performed by this combined system. In fact, Ethereum has stirred a lot of eagerness through the launch of its method for doing contracts. That allows the user to create their own account of cryptocurrency and use the same to make payment using Smart contracts. The company also allows the user to use ether, a cryptocurrency, in areas of governance, autonomous banks, financial derivatives, trading and settlement by using Smart contract system (Crosby et al. [3]). Rapidly changing external forces including technological infrastructure, social trends, drastically fizzling ecological environment and of course disintegrating sustainable economic indicators. Most important of these changes is increasing the population of the world but reduced sustenance from the natural resources to their rapidly changing economic activity. Every single person has several instantaneous commitments to meet; he wants to meet them all in the least possible time, to gain maximum benefit from the shortest possible time. This means that he should be calm and remain undisturbed through unwanted intrusions while staying at his residence. Which is for a very short time, owing to his overcommitted self, this greed for more power and control, through efficient use of resources. Natural resources no doubt are a necessity for performing economic activity nevertheless; they are limited and cannot sustain greed for power and control. Therefore, implying that socio-economic and depreciating ecological system are the forces that drive toward the development of smart cities.

According to Crosby et al. [3] Samsung, IBM, Citi and other notable names are all researching the uses of the Blockchain to make their work even more refined. Recently, many notable banks including Barclays, JP Morgan, State Street, UBS,

Royal Bank of Scotland, Credit Suisse, Commonwealth Bank of Australia and Gold-man Sachs. Have teamed up with a New York-based company going by the name of R3 the purpose of the group is to use Blockchain profitably and efficiently to better the performance of the financial market. Vistas for use of Blockchain are wide and once it is fully understood, it can be applied to any sector. NASDAQ is private equity-based firm providing cap table and investor relationship management for pre-IPO or private firms. The prevalent method depends on third parties efficiency thereby, making the system slow. Recently the company joined its hands with chain.com to implement a Blockchain backed system to bring efficiency into its services.

2   Requirements and Analysis of Smart City

A smart city is considered an epitome for the entire world in a certain field of study. By the Disruptor Daily [4] surveyed and found that increasing reliance on technology and increasing concentration of population toward better economic prospects. Shall result in almost 65% of the global population living in the city by 2040. The need for medical services, the security of personal data in the midst of rising technological use. Shortage of time to execute property transfers in a legal manner and risk-free method. Management of natural resources and public fund efficiently and effectively shall require transparent data. Because the magnitude of the concentration shall fuel the corruption within the government sector worth billions. Further, while being plagued with shortening natural resources and society's wish to sustain a global economy shall require transparency. Transparency of data at national level shall give rise to the culture of questioning, which in return shall help bring efficiency in the planning for the future goals of the economy. Blockchain can help give insight into the government's usage of public funds, while the use of sensors to gather information on public needs and conclusions on that data [4].

Sharma et al. [5] have observed that a Smart city means efficient usage of resources. Efficiency also means the least possible time as well as the use of input; therefore, the author suggests using Blockchain-based public transport system for the movement of commuters through the well-laid-out city. A traveler can access the publicly available online booking system without having to rely on banks for pay-ment to authority or risk of a payment being transferred to another person's account. Further, since, the Blockchain uses crypto (code/Hash) system information about the transaction and even the concerned parties of the contract are non-readable. This makes it a risk-free system that shall help secure personal information of the person on a technology-reliant system. For instance, corporate personnel travelling in China is visiting the company's manufacturing facility and then travel back to headquar-ter located in Europe with transport system dependent on Blockchain, the person can himself select the airplane and train and pay to relevant people directly. Sharma et al. [5] have explored the impacts of using Blockchain in smart cities. The system in these cities is linked with technology hence, the loss of data due to hackers impregnating central server is high. Usage of Blockchain can help avoid this mess because data shall be accessible by authorized public from several machines connected to the network. Further, tempering a single block shall instigate a series of changes in chain-linked several other blocks. This means that change is practically impossible because the information being presented to the user is in Hash form hence, making it impossible for the criminal to temper with the data. Further, the system can help bring in efficiency by benefiting the user from doing it right the first-time philosophy. Not only helping to overcome costs of error but also reduce the time taken by the third party to expedite a contract [5].

Rivera et al. [6] have concluded that for a smart city it is necessary that it stays on top of things on a continuous basis. Since, the city's success depends on the acceptance of global technological, social, economic, ecological and legal needs. Therefore, the government alone cannot comprehend these forces nor, it can make decisions based on unchallenged ideas. Use of Blockchain for making available information about future decision to sell electricity to the neighboring country can help arrive at a viable decision through fear of questioning. Politicians shall be afraid to receive kickbacks and cause loss to the national fund [6].

3 Blockchain for Smart Healthcare Applications

3.1 Motivation and Challenges

Current governments, hospitals and patients are concerned with the loss of personal data to the hackers of the system. People intent on breaking into private data have been able to breach even the well-secured system and sell the data to the corporate world for millions. Not only that, the personal information had been used to cause a rupture in the financial system, causing a dent in people's finances. A hospital is a place where the data is needed in order to treat the patient in a timely manner however keeping it safe is becoming a pain that is asking to be relieved. The need of data but, date brings problems with it and authors to find. Apart from data management collected from the patient in order to clear outstanding dues for the services gotten from the hospital. Data portraying the patient's health history is also very useful to the doctor on duty. Imagine a patient transferred from another hospital to a much-advanced facility remaining untreated because the doctor did not know about his previous history. The prior hospital in fear of losing the data did not maintain a systemized ledger and only the doctor who attended him knew about the patients' predicament [7]. This is not smart, the delay can cost the advanced city a man who might be a key figure in the industry the city is globally known for.

3.2   Reliable Healthcare Data System

Healthcare system is supposed to be efficient to treat the patient as per the disease diagnosed by the doctors. The system shall be considered best in the world if it is able to generate interest in the global community and they come to be treated in that country. As have been revealed above that by 2040 approximately 63–65% of the entire global population shall be living in cities by the Disruptor Daily. This means that the local governments shall have to revamp the current healthcare system manifold. Healthcare in the smart city also means that the services ought to be technologically advanced in order to get the sector recognized globally. To attain this not only the medical research should be finest at introducing cost-effective and efficient techniques that lead the global medical field. In addition, the methods should be able to treat the patient on the basis of doing it right the first time and in the least possible time. In such cities, doctors on duty should be able to treat the patient by just accessing the information about his health appearing in hospital records [8]. Apart from this, the system should also protect the personal information of the patient including debit/credit card details, identity and address.

3.3   Role of Blockchain in Healthcare System

Blockchain system can not only help avoid loss of critical personal patient data but save millions spent on securing it. The system uses crypto grapey, a technique that presents the information in algorithmic codes. This, crypto graphing of patients' personal identity can help avoid losses caused by hacker's attacks. Further, unlike the server system where data is stored only on one computer which may get corrupted hence, losing all the data stored on it since the patient's inception into the hospital. The Blockchain splits the data among several nodes connected to the network. Thereby, making the data accessible from several authorized machines, hence, reducing the risk of loss due to the system's corruption. Because data is sharable among the peer to peer connected systems [9] (Fig. 1).

3.4   Trusted Blockchain in Healthcare

According to Engelhardt [10], the Blockchain system can help alleviate the worries of the patient Fig. 2, knowing that his personal information in the internet-connected city is safe and not viewable. Further, the system shall also give timely information to the doctors to determine the disease of the patient, current level the disease is at. Additionally, in the smart city, the patient does not have to worry about going to a foreign country just because his government has not been able to come up with funds for sponsoring the research in the area that is plaguing the society [10].

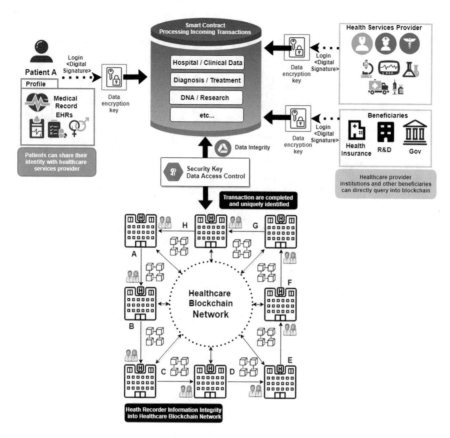

Fig. 1 Blockchain for intelligent healthcare system

Through the use of sensors and other devices, the government can gather data about the diseases that are plaguing society. Thereby giving it a chance to channel the required resources to come up with a cure for the identified disease. Not only this, the patient while being in bed can relay to the doctor through sensor what he is feeling at the moment. Helping doctors determine the pain and imbue patient with right injections causing a reduction in treatment cost. The usage of Blockchain according to the author can help reduce the costs of healthcare significantly via efficient usage of data regarding a patient's history.

4   Intelligent Vehicle Technologies

Intelligent Vehicle is the combination of various technologies which consists of LTE or Long-Term Evolution, GPS or Global Positioning System, OBD (On-Board Dige-nesis) and camera. Long-Term Evolution or LTE is a standard for wireless broadband

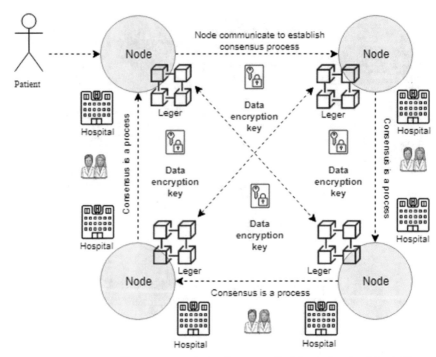

Fig. 2 Trusted blockchain environment in healthcare system

communication for data terminals and mobiles connections which is based on the GSM/EDGE and UMTS/HSPA technologies [11]. The main aim of LTE is to improve the efficiency and speed of wireless data networks by making use of new DSP or digital signal processing techniques and modulations that were developed around the thousands of years ago. And is also utilized to redesign and reduce complexity of the network architecture to an IP-based system with significantly lower the transfer latency compared to the 3G architecture. The LTE wireless interface is not compatible with 2G and 3G networks and thus it must be operated on a separate radio spectrum. Global Positioning System or GPS is a satellite-based radio navigation system possessed by the United States government and controlled by the United States Air Force. It is a global navigation satellite system and it provides time information and geo-location to a GPS receiver anywhere on or near the Earth where there is an unblocked view to four or more GPS satellites [12] (Fig. 3).

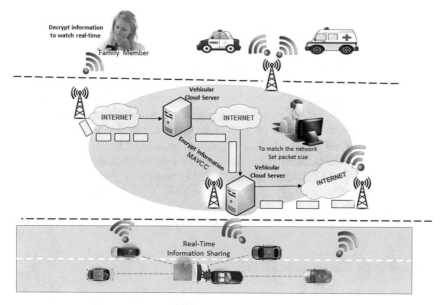

Fig. 3 Overview of intelligent vehicle technologies

4.1 Trusted Vehicle Networks

As per Eze et al. [13], recent development in the wireless technology and vehicle automation systems dependent on the sensors has laid the ground for research in VANET. The system consists of V2V (vehicle to vehicle) and V2I (vehicle to infrastructure) communication. In fact, Toyota the sixth-largest company in the world is making cars avoid accidents based on the information from other Toyota cars travelling in the area and the cars around the car itself. The system is one-off it's kind and unique, it is going to change the traffic flow system. As other cars may be able to inform the driver about road or route conditions ahead and the driver may take an alternative route. Further, the ITS is already being used in developed countries to help monitor emissions from the transport system. Manage a fleet of city buses or other intra-city transport methods according to passenger flow. In addition to that, the system has presented a reward based on Vehicle Blockchain communication technology for a specific services to remove centralize authority of communication between Vehicles to Vehicle [14].

4.2 Smart City Vehicle Challenges

A smart city can be known for its wide and well-connected transport system that is the key to help generate economic activity in a seamless manner. Economic activity depends on the on-time availability of human resources, the efficient transport system is affordable for the masses. Further, the spread of system across the entire city is necessary to enable the human resources to travel from a residential zone to different industrial zones that cover one side of the city. Nevertheless, these industrial zones are spread out and the transport system must manage itself according to the commuters' traffic. To achieve this, sensors and wireless alone cannot be enough to make the system work in harmony within a densely populated city. Because the city is home to private vehicles as well and the public transport has to contend with all these obstacles on the road to get all human resources on time. This means that the ITS shall also rely on the information being given to it by autonomous vehicles. Such vehicles are still in development and may need time until then a well synchronized ITS that can plan fleet management based on traffic on-road and the users is difficult to apply [15].

4.3   Role of Blockchain in ITS

ITS is relying on the information generated from the sensors, the current traffic management system does not help the public to be able to plan the route based on avoiding traffic jams. Blockchain can help available the data on the screen of the Wi-Fi connected car for the driver to see and plan accordingly. Further, the mass transport system in order to reduce ecological impacts on the environment is using alternative greener forms of energy. The energy usage is relying on data generated by the sensors and other appliances used in the electrical system to switch between inputs for energy. The smart grid system can help the generators of electricity change the load to the transport system according to commuters' requirement. Further, the system can switch between the inputs used to generate this electricity based on the information about resources availability. Use of Black chain can also help avoid blackouts caused due to loss of data from the central authority of the system or occurrence of the problem in the central authority of the system [16].

4.4   Impact of Blockchain in ITS

Yuan and Wang [17] observe that commuters using ride-hailing service or any other form of transport have to relinquish data to a third party. Who is managing payments on behalf of the transport service provider, further, the transporter is supposed to give him cut from payment? Via block-chain, this third party is cut out and the passenger shall be able to communicate to service giver and pay him directly as well. Further, currently due to centralized authority controlling the vehicular information public cannot plan the journey and avoid the traffic jams. Usage of Blockchain shall enable this data to the Wi-Fi-connected car drivers helping travelers save fuel and cut down emissions as well. One thing more, an efficient transport system is based on public transport fleet management, use of Blockchain can help predict consumers to switch between the means of transport. They can do this based on the availability of information about the load on the mode's fleet helping them save their time [17].

5 Disseminate Self-governing Organization

Kahle and Stulz [18] have found that financial crises have occurred because of the use of complex financial models and lack of transparency of actual transactions occurring in the corporate circles. The mess is also fueled by the public's lack of financial knowledge and quick accountability effects for dishonest reporting. Similarly, corruption in the governance system is also due to a lack of visibility on the inside doings to the public. This is the reason that expensive government projects face cost overrun because the public cannot question while not knowing.

Mirabeau and Maguire [19] point out that the shareholders and not the professionals sway an autonomous organization. This results in the transparency of the inside doings of the organization through questions. Although, questioning about decisions brings inefficiency in the form of delayed decisions. Yet, fear of accountability shall reduce the chances of fraud or inefficient usage of resources. This brings efficiency because the shareholders are the external forces hence; they require the value and therefore, know better what is needed by the masses [19].

6 Blockchain in Smart Education System

As per Gibbs and Jenkins [20], UKs' education system is world-renowned most of its universities every year make it to the top and host a second home to global students. This has been achieved because of a persistent focus on the quality of the education system by trial and testing different techniques. This is the reason that the methods adopted by this system are replicated in other countries. However, the concentration of global students toward the UK has created a myriad of problems for the UK government as well as the students. Online registration with the university requires personal information of the student. This registration often happens through the third party, resulting in the stealing of information by people with negative intents. The fraud not only causes financial problems for the students but also give a bad name to the University for ignoring to streamline its registration process.

Another problem is the lack of data available about students from local countries, countries that have suffered racial and violent terrorism. This lack of information about the students sometimes results in racial discrimination instigated against them by local students. Racial violence in the universities is a reality, not only the difference between the black and white are a cause of major concerns for the university administration. Simultaneously, verbal abuses against minorities is also a reality and people belonging to other religion often have to face monkey chants. Sometimes followed by a brawl among students [21].

Currently, there are only a few selected universities available throughout the world that are disseminating the best education for several fields of study. This has benefits as well as negative consequences, country playing host to foreign students get to benefit from the increased taxes, earned through fees earned by universities from these students. Such countries UK and USA in particular also get a chance to avail a more talented and much younger work-force to replace their ageing people in the production sector. Thereby, retaining people of creativity linked with local culture and honed by US and UK education to become revered scientists, doctors, accountants, and engineers. On the other hand, the country from where the students come has to pay through foreign reserves. UK and US universities do not accept local currency this forex causes a loss to trade balance of the economy of that country. A country that is already reeling back from the red figure trade balance is forced deep by the expensive higher education. The problem becomes further exacerbated when the student does not return to serve his own country through advanced education learned in foreign universities.

Osipian [22] that usage of Blockchain in the smart city-based education system shall help reduce the chances of educational frauds being committed by the third party giving a student an assurance for admission. Through well-connected peer network using smart contract, the university fees and legal registration method shall become very easy. Third-party involved in helping students register is going to be eliminated from the chain. Students after viewing online available information shall be able to pay directly to the University of their choice, reducing the cost to them and university itself. On the other hand, information about the students shall be secured through the Blockchain crypto graphed data system, making it non-viewable.

Sun Yan and Zhang [23] have observed that Smart city is part of the center of the global economy in the way that it can do best. London being home to the highest number of best-ranked universities is the only part of the globe that has a cluster of best education providers [21]. This means that it is a smart city because it is disseminating higher education that is revered by the global community. It is still leading the world of education by making other people replicate what it can do best. Using Blockchain and a smart contract shall make it even much smarter by securing student's personal information and registration with the University of Student's Choice via smart contract. The system shall help reduce the registration cost and time of the student by making sure that the university gets the fees directly and not through somebody else. Thereby, increasing the revenue, which otherwise is reduced by middle agents fees. On the other hand, shortening of registration of process and wide availability of information shall also increase the number of admissions.

7 Decentralized Distributed Insurance for Smart Living

By Elmaghra [9] have found that living smartly means that everything that the citizens use is connected to the internet through the Internet of Things. Yes, the world is not moving toward full connectivity, if automakers are able to control cars through wireless devices then why not homes cannot be controlled in that way. A person coming from office is able to find his home's temperature cozy in whatever weather because the sensor in the house is connected with a person's smartphone. The phone has a GPS locator tracking his movement, however, life has become easy and yet, a person's privacy has been obliterated entirely, helped by the easy availability of his personal information. Websites too are collecting the information in exchange for their services, this, happens when the user is trying to learn something educational. This information is not coded and hence can be viewed by the buyer of the information from the website owner. This lack of privacy has laid him bare for unfriendly eyes, yes; these people have also become smart and can control sensors by using wireless technology. This negative use of technology is a threat to persons' life hence, the need for insurance against this risk. Bauer et al. [24], describes that insurance is a system that covers the risk that might arise in the registered person's life. Risk is different from the scenario in question, that is to say, insurance can provide coverage to a person's life. It can also validate the risk to a personal tangible loss. Current insurance system is very cumbersome; a person has to rely on the services of other people, costing him time and money. Similarly, an insurance company has to relinquish part of its revenue in the form of fees of the professional is hired to evaluate the damage inflicted on the registered person due to the fulfilment of a condition. There is another problem; independent verifier of the situation may be scheming with a rival group. Since the person is well connected with authorities this might then taint the results of examination against the registered person. Thereby, losing him the legitimate benefit. The cost-effectiveness of the current insurance system. A person is registering for insurance with party however; the payment is made through the bank. This method is secure because the registered person is getting feedback regarding payment from an independent third party that is being managed by a legal government entity. On the other hand, in circumstances under which the situation specified in the contract arises then the registered person has to wait for the results of the analysis performed by the third person. Consuming a lot of time of registered person and affecting his daily routine (Fig. 4).

According to Crawford [8], use of Blockchain technology in a smart city-based insurance system can help protect a person's private information because data is chain-linked crypto graphed. Yes, Blockchain uses peer connected network and though public can access it nevertheless, data cannot be copied unless authorized by all the nodes of the system. Moreover, the data is not centered in one server, rather, it is scattered on all the nodes on the system. Additionally, by using smart contracts insurance company can receive direct payments from the client in the form of digital currency. Helping reduce the cost of bank's service charges, if this is not

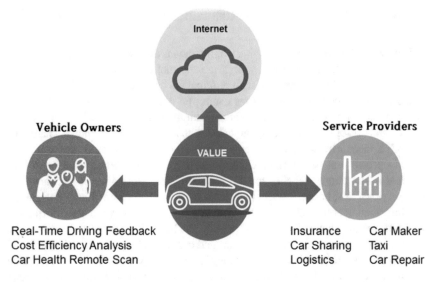

Fig. 4 An overview of intelligent connective for smart city

enough, since the live information regarding arising of conditions is available via sensors this deletes the need of third person's verification. Thereby, reducing the chances of unfair decisions.

8 Distributed Smart Real Estate

We have found that real estate is a platform that is swayed by developers and the government. From zoning to the registration of property, from payment of taxes to the creation of contracts and development of necessary documents of verification. Third-party currently perform these services while the cumbersome registration and tax structure are also centralized and a cause of inconvenience to the public as well as an unnecessary dent in the government's finances. The developers also do not take into account the economic indicators of the society hence, suffering the lower and middle classes. The loss of time due to the reliance of the public on the developers for the creation of legal documents not only cuts in the government's finances. However, also loses public their precious time ensuing in the tense environment at the workplace, causing the economy to suffer. Therefore Veuger [25] gives an opinion on Blockchain and real estate sector to the next level the wide availability of information and transparency on financial transactions. Shall help nip in the tax-avoiding tactics adopted by millionaires, corporations and heavyweight individuals. Because the use of Blockchain shall make the data available to government authorities. Who are able to transact because of lack of transparency in the banking system. This is inefficient for a smart city aspiring to lead the world because of the system affects the government's revenues. Use of Blockchain in the real estate sector in addition to the smart contract shall make available the data on property widely available. This transparency shall help bring in public's efficiency; the public currently has to rely on banks and lawyers for payments or legitimacy of the documents. This is inefficiency and a smart city is about bringing in efficiency by cutting out unproductive time. The possible use of Blockchain and Smart Contracts in the real estate sector. Data about the property in question shall be available on the government website and the buyer will simply be able to verify documents truthfulness. Further, using smart contracts shall enable direct payment transfer into parties account via cryptocurrency at the time the system deems that conditions have been met. The usage shall put an end to the cumbersome escrow and other services. Further, buyers' worth shall remain anonymous in this way and this mystery shall help reduce the intention of crime otherwise fueled by the clarity about buyers' worth [25].

9   Sustainable Green Technology

Dornfeld [26], have found that green technology is the effect of the research on the finding of an alternative to a non-renewable resource. Purpose of this technology is manifold; reduce the impact of human activity on the environment of this world because its disintegration is causing a massive climate and aquatic change. Further, use the input resources in an efficient manner less input more output. The concept is based on the logic of profitability in long-term; revenue over cost shall result in a positive balance, leaving behind a net to sustain similar activity in the future.

As have been exclaimed above, smart city definition evolves with each new iteration of an efficient and effective manner that is making a life of citizens cleaner and less explosive toward the environment. Khan et al. [27] found that micro grid is an efficient system that can help in a city becoming the cleanest place of residence while simultaneously turning into a global hub of economic activity. However, the system has many challenges to overcome, first is the availability of data regarding energy usage from each plot in the city. Next is the integration of this data into EMS of the micro grid system, third, since micro grid is the sum of technical parts, therefore, it makes it necessary for feeling its full-fledged effects that other part makers are technically advanced as well. Ottman [28] expounds and finds that sustainable strategic direction can only be achieved if the economy as a whole consider SROI. The figure is a return on investment albeit added with the social impact of an activity. The percentage can help arrive at the closeness of strategic direction with ecological, economic, social, legal and technological factors of changes. Sun Yan and Zhang [23] have observed that efficiency and effectiveness have no clear definition. Blockchain system can help reduce time and at the same time bring the security of data and direct dealing between parties to a contract. Most output generated in limited time, reduction in fraud due to the directness of transaction and non-mutability of data is pre-requisite for a sustainable green technology-based smart city. Since because re-performing has its own cost and limited natural resources cannot afford non-value generating cost.

10 Conclusions

This chapter discussed the Blockchain technology adoption perspective prevailing in smart cities applications. Blockchain technology is built on distributed ledger technology which allows data entered into the system to be fanned out amongst all users. Today Blockchain technology has been applied in many areas such as banking, healthcare, education, and commerce. The use of Blockchain cannot only help the public in dodging the clogged roads by viewing the current traffic on the roads ahead. But also switch between the modes of transport to reach the point of destination in time based on the estimated time remaining in the arrival of the next vehicle. The personal information of passenger can be secured through Blockchain and is received by service giver reducing the unfair usage of information. Further, usage of personal cars may be lessened based on the widely available information to the public about public transport system thereby, helping to reduce ecological and fuel costs. However, this chapter focused on Blockchain potential explored how it can be used to improve the current smart cities application. The overall of Blockchain advantages of giving some of the Blockchain technology-based smart city applications, challenges, and future opportunities.

References

1. Blobba et al (2017) Smart City Hub. Available. https://smartcityhub.com/governance-economy/smart-city-smart-story/. Accessed 7 Aug 2019
2. Albino V, Berardi U, Dangelico RM (2015) Smart cities: definitions, dimensions, performance, and initiatives. J Urb Technol 22(1):3–21
3. Crosby M, Pattanayak P, Verma S, Kalyanaraman V (2016) Blockchain technology: beyond bitcoin. Appl Innov 2(6–10):71
4. Blockchain for Smart Cities: 12 Possible Use Cases. https://www.disruptordaily.com/blockchain-use-cases-smart-cities/. Accessed 20 Oct 2019
5. Sharma PK, Moon SY, Park JH (2017) Block-VN: a distributed blockchain-based vehicular network architecture in smart city. JIPS 13(1):184–195
6. Rivera R, Robledo JG, Larios VM, Avalos JM (2017) How digital identity on the blockchain can contribute to a smart city environment. In: 2017 International Smart Cities Conference (ISC2), pp 1–4
7. Singh D (2011) Future internet services for e-healthcare monitoring applications. VDM Verlag Dr. Müller Publisher, Germany, p 172. ISBN: 978-3-639-36608-2
8. Crawford M (2017) The insurance implications of blockchain. Risk Manage 64(2):24
9. Elmaghraby AS, Losavio MM (2014) Cybersecurity challenges in Smart Cities: safety, security and privacy. J Adv Res 5(4):491–497
10. Engelhardt MA (2017) Hitching healthcare to the chain: an introduction to blockchain technology in the healthcare sector. Technol Innov Manage Rev 7(10)

11. Singh M, Kim S (2018) Trust Bit: reward-based intelligent vehicles communication using blockchain. In: The 4th IEEE World Forum on the Intelligent of Things (WF-IoT), Singapore

12. Guerrero-Ibanez JA, Zeadally S, Contreras-Castillo J (2015) Integration challenges of intelligent transportation systems with the connected vehicle, cloud computing, and internet of things technologies. IEEE Wirel Commun 22(6):122–128

13. Eze EC, Zhang S, Liu E (2014) Vehicular Ad Hoc Networks (VANETs): current state, challenges, potentials and way forward. In 2014 20th international conference on automation and computing. IEEE, pp 176–181

14. Singh M, Kim S (2018) Crypto Trust Point (cTp) for secure data sharing among intelligent vehicles. In: The 2018 international conference on electronics, information and communication (ICEIC 2018), Sheraton Waikiki Hotel, Honolulu, Hawaii, USA, 24–27 Jan 2018

15. Hofmann E, Rüsch M (2017) Industry 4.0 and the current status as well as future prospects on logistics. Comput Ind 89:23–34

16. Singh D, Singh M, Singh I, Lee H-J (2015) Secure and reliable cloud networks for smart transportation services. In: The 17th IEEE international conference on advanced communication technology (ICACT2015), Phonix Park, South Korea, pp 358–362, 1–3 July 2015

17. Yuan Y, Wang FY (2016) Towards blockchain-based intelligent transportation systems. In: 2016 IEEE 19th international conference on intelligent transportation systems (ITSC). IEEE, pp 2663–2668

18. Kahle KM, Stulz RM (2013) Access to capital, investment, and financial crisis. J Financ Econ 110(2):280–299

19. Mirabeau L, Maguire S (2014) From autonomous strategic behaviour to emergent strategy. Strateg Manag J 35(8):1202–1229

20. Gibbs G, Jenkins A (2014) Teaching large classes in higher education: how to maintain quality with reduced resources. Routledge, London

21. The Guardian (2019) A demeaning environment: stories of racism in UK universities. Available at https://www.theguardian.com/education/2019/jul/05/a-demeaning-environment-stories-of-racism-in-uk-universities. Accessed 9 Aug 2019

22. Osipian AL (2014) Will bribery and fraud converge? Comparative corruption in higher education in Russia and the USA. Compare: J Comp Int Educ 44(2):252–273

23. Sun J, Yan J, Zhang KZ (2016) Blockchain-based sharing services: what blockchain technology can contribute to smart cities? Financ Innov 2(1):26

24. Bauer AR, Burns KT, Esposito MV, O'Malley PL, Olexa BJ, McMillan RJ, Progressive Casualty Insurance Co. (2013) Monitoring system for determining and communicating the cost of insurance. U.S. Patent 8,595,034

25. Veuger J (2018) Trust in a viable real estate economy with disruption and blockchain. Facilities 36(1/2):103–120

26. Dornfeld DA (2014) Moving towards green and sustainable manufacturing. Int J Precision Eng Manuf-Green Technol 1(1):63–66

27. Khan S, Paul D, Momtahan P, Aloqaily M (2018) Artificial intelligence framework for smart city microgrids: state of the art, challenges, and opportunities. In: 2018 Third international conference on fog and mobile edge computing (FMEC). IEEE, pp 283–288

28. Ottman J (2017) The new rules of green marketing: strategies, tools, and inspiration for sustainable branding. Routledge, London

Chain of Antichains: An Efficient and Secure Distributed Ledger

Jinwook Lee and Paul Moon Sub Choi

Abstract Since the inception of blockchain and Bitcoin (Nakamoto, Bitcoin: A peer-to-peer electronic cash system (2008) [18]), a decentralized-distributed ledger system and its associated cryptocurrency, respectively, the world has witnessed a slew of newer adaptations and applications. Although the original distributed ledger technology of blockchain is deemed secure and decentralized, the confirmation of transactions is inefficient by design. Recently adopted, some distributed ledgers based on a directed acyclic graph validate transactions efficiently without the physically and environmentally costly building process of blocks (Lerner, Dagcoin: a crytocurrency without blocks (2015) [17]). However, centrally-controlled confirmation against the odds of multiple validation disqualifies that newer system as a decentralized-distributed ledger. In this regard, we introduce an innovative distributed ledger system by reconstructing a chain of antichains based on a given partially ordered pool of transactions. Each antichain contains distinct nodes whose approved transactions are recursively validated by subsequently augmenting nodes. The boxer node closes the box and keeps the hash of all transactions confirmed by the box-genesis node. Designation of boxers and box-geneses is conditionally randomized for decentralization. The boxes are serially concatenated with recursive confirmation without incurring the cost of box generation. Rewards are paid to the contributing nodes of the ecosystem whose trust is built on the doubly-secure protocol of confirmation. A value-preserving medium of payment is among numerous practical applications discussed herein.

Keywords Blockchain · Distributed ledger technology · Decentralization · Antichain · Partially ordered set · Consensus protocol · Stablecoin · Cryptocurrency

J. Lee
LeBow College of Business, Drexel University, Philadelphia, PA 19104, USA
e-mail: jl3539@drexel.edu

P. M. S. Choi (✉)
College of Business Administration, Ewha Womans University,
Seoul 03760, Republic of Korea
e-mail: paul.choi@ewha.ac.kr

© Springer Nature Singapore Pte Ltd. 2020
D. Singh and N. S. Rajput (eds.), *Blockchain Technology for Smart Cities*,
Blockchain Technologies, https://doi.org/10.1007/978-981-15-2205-5_2

1 Introduction and Our Motivation

The original vision of the Internet is to construct a global system of interconnected computer networks. The Internet has revolutionized the modern society and prosaic activities—we became constantly connected. Yet, even though we are in the era of shared economies with help of the Internet, the individual service providers and their central authorities are not as much mutually-beneficial as they can be. For example, the drivers of shared rides and the hosts of shared accommodation can be better-compensated with more decentralized discretion. The central authorities of those industries can streamline the ever-increasing costs of cyber-security by decentralizing their centralized systems with the advent of decentralized-distributed networks. Decentralization clearly synergizes all stakeholders of the emerging value chains of shared economies, regardless of the size and side of transactions. The conventional centralized mechanism can be decentralized using the distributed ledger technology (DLT), more colloquially referred as the blockchain technology.

Upon the creation of blockchain by Nakamoto [18], the DLT has gained dramatic attention over the last decade among developers, practitioners, and academics alike. The unprecedentedly multiplied market values of cryptocurrencies, including blockchain-associated Bitcoin, have inspired a bevy of practitioners and researchers to implement a variety of real-life applications on distributed peer-to-peer (P2P) network systems. The original blockchain is considered decentralized and secure with recursively serial confirmation. However, the environmentally and physically arduous process of block building ("mining") is the root cause of systemic inefficiency. This causes ramification ("forking") of cryptocurrencies beyond Bitcoin. Recently, directed acyclic graph (DAG)-based distributed ledgers—including Hashgraph [1], IOTA [25], Byteball [3], etc.—have emerged as a new generation of DLTs. A DAG bears the shape of an irregularly tangled leaf with recursively validating, binomial sub-trees, and it is computationally efficient without mining. However, a DAG-based DLT typically requires centrally-managed nodes to perform the final confirmation of transactions and, to the worse, suffers from the odds of double spending.

Our approach to DLT is synthetic: Combine the decentralized and secure features of blockchain with the efficiency of DAGs. An antichain ("box") is a set of distinct, unrelated elements ("nodes") that orthogonally complements a chain. As a DAG evolves in the original ledger, we synchronously reconstruct a dual ledger with a chain of antichains ("boxchain") as a serial concatenation of boxes of nodes that originate from the underlying DAG. Because a node in the DAG-ledger recursively approves up to two transactions of two previous nodes, a reflecting box in the dual ledger, in effect, recursively confirms the transactions of the last box. This doubly-secure consensus protocol is as follows: First, a new node in a box immediately verifies and validates the approved transactions of the previous node in the box. Second, when the last node in the box ("boxer," a light node[1]) is determined the head node of the

---

[1] A lightweight node (or light client) only references the full node's copy with less required memory capacity and processing power.

box ("box-genesis," a full node[2]) is randomly chosen among good-standing nodes other than the boxer. Third, the box-genesis finally confirms all validated transactions within the box and keeps the hash of confirmed transactions, followed by timestamp synchronization with all the previous box-geneses. This means that the box-geneses must have a copy of the entire network. Such timestamp synchronization also takes place to boxers, which only keeps hash pointers. At the core of trust, or governance, on the boxchain ecosystem, this consensus protocol is numerically impossible for a malignant intention of delayed or multiply-validated transactions. While the original DAG-ledger and the dual ledger synchronize and expand, there is no mining game involved contributing to the efficiency of the system. In sum, our dual ledger-keeping algorithm is decentralized, efficient and secure.

Although currently available P2P payment systems appear more facile and expeditious than before, they can improve significantly with embracement of DLT as both share similarities in chains of digital signatures of asset transfers and P2P networking, etc. In this sense, e-commerce marketplaces can be an ideal place to implement the idea and application of DLT. However, a cryptocurrency with volatile market value is not suitable for usage in payment, and in order for it to serve as a medium of exchange or a store of value, its value has to be stable or pegged against a fiat currency. In these regards, our DLT provides two distinct cryptocurrencies: boxdollar, a value-preserving medium of payment or store of value ("stablecoin"); and boxcoin, a crypto asset. Such a dual-currency system is desirable in sustaining a decentralized-distributed ledger network with proper incentives as well as expanding the ecosystem for diverse usages of stablecoins. In case of international payments for merchandise purchases, first and foremost is the highest reliability and stability—a proper digital currency must be able to function like a fiat money as a useful medium of exchange and a dependable store of value. Thus, pegging boxdollar to a fiat money (e.g., the U.S. dollar) is straightforward and intuitive. In order to maintain the reliability and stability of boxdollar, the optimized rebalancing of currency holdings per various criteria, e.g., [12] "safety first model" or conditional value-at-risk model, can be considered. Besides, boxdollar can be used as a local currency in a municipality with stable tax revenues. As the local governments with local stablecoins mutually agree to render them compatible, these value-preserving cryptocurrencies may eventually act as quasi-fiat currencies.[3]

While we need a stablecoin like boxdollar to finance purchases and various transactions, we also view the distributed ledger system as a society of healthy extent of "greed." It is not difficult to see that concerns may be overdone if a society only consists of good-standing, law-abiding participants. However, tantamount rewards paid out in boxcoin, a crypto asset, are essential for the peers with good-standing track records. In order to encourage rewarding behaviors, a fair opportunity will be given to every peer to play a vital role in the system. In our ecosystem, there are two

---

[2]Since the mining process (i.e., solving a partial hash inversion problem) is not required in this system, prohibitive computing equipment is not necessary for the full node, which only requires sufficient data storage capacity.

[3]See Sect. 4 for the discussions and models of boxdollar.

particular positions—the boxer and the box-genesis—to which any qualified participants can be designated multiple times. Besides, small fees will be charged to issued
transactions to provide more security to the system. Without fees, there can be nonsensical transactions burdening the efficiency and security of the system. The amount
of fees will be from negligible to small, depending on the matter.[4] The mentioned
elements such as rewards and fees must be realized instantaneously upon occurrence
of events—giving prompt incentive and motivation to the peers. When the ecosystem
gets larger, more robust and multiplied, it is obvious that the value of its currency
will gain given a fixed quantity of currency in circulation. This possibility of capital
appreciation is a great incentive for the peers to abide by the code of conduct. Given
these accounts, we are motivated to build a new distributed peer-to-peer system with
two distinct cryptocurrencies: boxdollar and boxcoin. Boxdollar is a stablecoin for
our daily use while boxcoin is a crypto asset to improve and manage our ecosystem
along with all participants' active engagement.

The organization of this research is as follows: Sect. 2 introduces our new idea
of the dual ledger-keeping algorithm, which is based on decomposition of partially
ordered sets (DAGs). Based on the primal layer of a DAG-based ledger, a dual-ledger
of boxchain (chain of antichains) is synchronized in the dual layer. A comparison with
existing DLTs and applications is provided. In Sect. 3 we present the mathematics of
our DLT: the nonhomogeneous Poisson process, combination of multiple transaction
flows, and the compound Poisson distribution. Section 4 provides the models of
boxdollar and discusses its potentials in the value chain, followed by our concluding
remarks in Sect. 5.

2 The Dual Ledger-Keeping Algorithm for Constructing an Effective and Secure Distributed Ledger

2.1 DAG: Distributed System, Differences from Blockchain, and Previous Works

A DAG, or acyclic digraph, consists of two entities: vertices and edges, i.e.,
$\mathcal{G} = (\mathcal{V}, \mathcal{E})$, which consists of a set $\mathcal{V}$ of vertices (or nodes) and a set $\mathcal{E}$ of edges
(or directed arcs). An edge is an ordered pair (i, j) meaning outgoing from vertex i,
incoming to vertex j, a pair (j, i) means from vertex j to vertex i. A DAG is deemed
an generalized blockchain because the graph itself is a ledger for storing transactions
where each vertex is intrinsically a single block. Note that in this section we use
the term DAG and $\mathcal{G}$ interchangeably. Besides, the terms "vertex" and "node" are
equivalent—"vertex" will be used when we are more focused on graph itself and
"node" when transaction processes and their related topics are discussed. Some of
the notable differences of a DAG from the original blockchain ideas are as follows:

---

[4]See Sect. 2.4.3 for more details of the incentive system.

- The main structure of a blockchain is a single chain that consists of blocks (a chain of blocks) and each block is a set of multiple transactions. If we look at each transaction as a single block, it is not difficult to understand that such blocks and their connections can be depicted as a DAG. A DAG can also be expressed in terms of trees—note that if there are N vertices (or nodes), then there can be *at least N* binary trees (see Sect. 2.2).
- Blockchain is based on synchronous time stamps, while a DAG is built asynchronously operating. However, our dual-ledger system is updated in a nearly synchronous manner (see Sect. 2.4).
- In a DAG there is none to force to separate participants into different categories. On the other hand, Bitcoin and some other cryptocurrencies justify to maintain two separate types of participants, required either to issue or to approve transactions. There are no "miners" creating blocks and receiving rewards in DAG.

Let us briefly mention how a DAG system works as a distributed ledger. Transactions are issued by nodes and their edges show what previous transactions they approved—users need to approve previous transactions in order to issue their own transaction. All transactions have its own weight, and in our case they are all equal to one which is adaptable as the system evolves. Speaking of the approval and the weights, the validation comes with cumulative addition of weights and it goes to all the connected previous nodes (i.e., ancestors of the node). Unapproved nodes in a DAG are called "tips." See Fig. 1.

A transaction validation process is simple—A node selects some tips at random and verify their validity in order to issue its own transaction. It is assumed that the approving nodes checks whether or not the two connected transactions are conflicting in a diligent and honest manner. Note that this validation is different from the final confirmation in terms of consensus mechanism. A node even before the first transaction is called the "genesis." The genesis node has all tokens created in the beginning of the DAG (no more tokens will be created) and then send the tokens out to several founder nodes. In order to issue a transaction a node needs to approve the two tips and solve a cryptographic puzzle (i.e., a hash inversion problem) similar to those in the Bitcoin blockchain. For more technical details we refer the reader to [3, 11, 17, 19], etc. [3] and [25] investigate many important deterministic and stochastic aspects of the network flows on DAGs, respectively.. We will discuss both deterministic and stochastic aspects here for improvements of the framework.

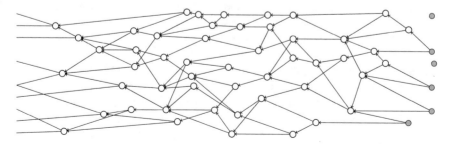

Fig. 1 Transactions on a DAG. Unapproved, gray nodes are tips

2.2 DAG, Structure of Trees and Related Set Representation

We continue to develop and modify the existing cryptocurrency models on the DAG, and more importantly, how we make the connection between the blockchain and a DAG. We will begin with an introduction of basic notions and develop such concepts using the system of distinct representatives, i.e., decomposition of a partially ordered set (poset). A simultaneous creation of an efficient network layer (called the dual-layer) will follow, and this is the core of our dual-ledger system.

Our space is a DAG $\mathcal{G}$, where vertices (or nodes) designate transactions and edges (or directed arcs) denote how they are connected—the edge (i, j) means that node i approves node j. In our network system a node selects exactly "two" tips at random and verify their validity in order to issue its own transaction. More details are presented after this section. There is another notation we need to define together with $\mathcal{G}$. $\mathcal{N}$ designates the network associated with $\mathcal{G}$, which consists of sub-networks $\mathcal{N} = \{\mathcal{N}_1, \ldots, \mathcal{N}_V\}$, where $V = |\mathcal{V}|$. Thus, each network contains a subset of $\mathcal{G}$ where a large number of vertices can be included. However complicated these networks look, there are always starting and ending points, and these are the latest vertex (the source node) and the genesis (the sink node), respectively.

A DAG is a combination of trees which consist of child nodes and their parent and ancestor nodes. For a tree, *Parent* means the predecessor of a node; *Child* is any successor of a node; *Siblings* are a pair of nodes that have the same parent; *Ancestor* is the set of predecessor; *Descendant* is the set of successors; *Subtree* denotes a node with its descendants.

Remark 1 (Binary Trees) As a node is supposed to approve two previous transactions, it is a binary tree. It is easy to observe that the whole system can be separated into multiple binary trees. (There at least V binary trees if the total number of vertices is denoted by V.)

In Sect. 2.1 we briefly mentioned the basic mechanics of issuing and approving transactions in a DAG. In connection with trees, if a node v approves two previous transactions then such nodes that issued the approved transactions become parent-nodes (or parents) of node v. Note that this validation of the parents automatically approves the parents of that parents, followed by such recursion all the way to the genesis. If we select a single parent node out of two parent nodes then it forms a chain, i.e, a linearly linked, totally ordered set. This is very efficient and is regarded as one of the main benefits to manage transactions in a DAG.

However, there are a typically large number of such chains, which are only partially ordered as there are incomparable ancestors in terms of subset relation, or validation relation. A DAG is a poset itself. The multiple chains of a DAG is the main reason why the system is asynchronous, which makes it almost impossible for all peers to agree to a single version of the truth. This means the final confirmation from a reasonable consensus mechanism would be unreachable if the network system is solely based in a DAG. This motivates our dual ledger-keeping algorithm.

Remark 2 A DAG is a poset by subset relation.

Now let us turn our attention to see what happens to a DAG when a node is approved in terms of the cumulative weight.

Remark 3 (Cumulative weight on a node, a counting measure of integrity and trust) As mentioned earlier, we assume all nodes have their own weight equal to one. In addition to the references to the parents (by hashes), the cumulative weight is a good measure for the integrity and trust. By the validation of a single child node, its all ancestors' weights will be added exactly by one, which is convenient as a counting measure. In other words, the weight of a node provides the system with information of the number of child nodes at the moment. The exact cumulative weight of a node is not known to its neighbors due to the asynchronous nature of a DAG-based network.

Remark 4 (Inclusion of transaction information on nodes) Note that a child node encompasses its parent node by referencing its parent's hash. All information of ancestors (including the genesis) can be obtained (reference does not take any memory, only point to the location) in a recursive manner, which means that a child node has more information than its parent nodes.

Speaking of the information inclusion by nodes let us use the following notations:

$$T_k = \{\text{transaction information of node } i_k\}.$$

Suppose that node i_7 approved node i_6 and node i_5, and node i_6 approved node i_4 and i_3, and node i_5 approved nodes i_2 and i_1 as in Fig. 2. Then node i_7 can get the information of all 6 previous nodes and itself (of course), node i_6 includes nodes i_4, i_3 and itself, and node i_5 includes nodes i_2, i_1 and itself. For $k = 1, \ldots, 7$, let the set A_k designate the union of all ancestor nodes from the point of view of node i_k. This example can be written up as:

$$A_7 = T_1 \cup T_2 \cup \cdots \cup T_7,$$

$$A_6 = T_6 \cup T_4 \cup T_3, \quad A_5 = T_5 \cup T_2 \cup T_1$$

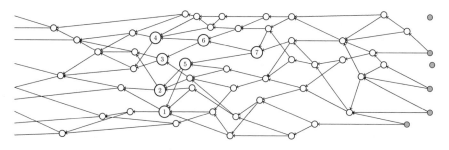

Fig. 2 Transactions on a DAG. Unapproved, gray nodes are tips (Compare it with Fig. 3)

and

$$A_1 = T_1, A_2 = T_2, A_3 = T_3, A_4 = T_4.$$

Thus, we have

$$A_7 = T_7 \cup A_6 \cup A_5,$$

followed by the subset relation

$$A_7 \supseteq A_6, A_7 \supseteq A_5, A_6 \supseteq A_4, A_6 \supseteq A_3, A_5 \supseteq A_2, A_5 \supseteq A_1, \tag{1}$$

$A_5 \not\supseteq A_6$ and $A_5 \not\subseteq A_6$ as they are only partially ordered. This is a simplistic example, but illustrates some important ordering on components of the network. Refer to Fig. 3, where child and parent nodes are upside down. In a tree graph, it is typical to have child nodes below, but in our special validation operation child nodes approve the transactions of their parent nodes. Again, the direction of edges are not from the genesis, but towards it.

Remark 5 (Not a typical binary tree structure) Again, note that the hierarchical relationship in a typical tree has ancestors above their predecessors. Our case is the opposite due to the selection-approval operation by a child node. This means, the root of a tree is a common child (the youngest "single" one!) of all the others. A child is a "superset" of the parent nodes.

Desides, the tree in Fig. 3 can be divided into three binary trees (i.e., repeated structural forms as illustrated), which will help solve a variety of problems in a recursive or iterative manner. We employ such recursion for our consensus protocol (in Sect. 2.4.1).

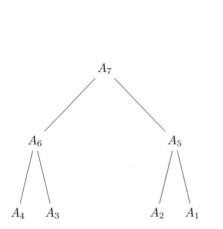
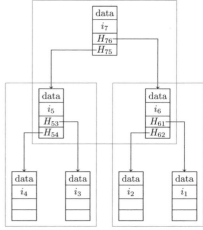

(a) Hasse diagram on partial order by subset relation.

(b) A set of binary trees with referencing hash functions.

Fig. 3 Inclusion of information in a subset of a DAG of Fig. 2

2.3 Primal and Dual Spaces, and the Dual Ledger-Keeping System

2.3.1 Partially Ordered Set and Its Related Concepts

In this section we present models and related algorithms making the system more integral, secure and efficient. We provide some basic definitions and results in connection with posets.

We say that two elements x and y of S are *comparable* if $x \leq y$ or $y \leq x$, otherwise x and y are *incomparable*. If all pairs of elements are comparable then S is *totally ordered* with respect to $\leq$. An element M of S is called a *maximal* element in S if there exist no $x \in S$ such that $M \leq x$ (i.e., $M \leq x \rightarrow M = x$). An element m of S is called a *minimal* element in S if there exist no $y \in S$ such that $y \leq m$ (i.e., $y \leq m \rightarrow m = y$). An element G of S is called a *greatest* element in S if $x \leq G$ for all elements x of S, and the term *least* element is defined dually. A *chain* (or totally ordered set or linearly ordered set) in a poset S is a subset $C \subseteq S$ such that any two elements in C are comparable (A chain is a sequence). An *antichain* in a poset S is a subset $A \subseteq S$ such that "no" two elements in A are comparable.

Remark 6 (Blockchain is serial, thus totally ordered, while a DAG is only partially ordered.) The main structure of a blockchain is a series of blocks—it is a serial concatenation of blocks, hence a totally ordered set. A DAG is only partially ordered and a mixture of piece-wise serial networks, seemingly disorganized. A DAG can also be restructured in a meaningful way by decomposing a poset into chains and antichains.

It is known that if every chain of a poset S has an upper bound in S, then S contains at least one maximal element (Zorn's lemma). If $x, y \in S$, then we say that y *covers* x or x is covered by y, denoted $x \lessdot y$ or $y \gtrdot x$, if $x < y$ and no element $u \in S$ satisfies $x < u < y$. The chain C of S is maximal if it is not contained in a larger chain of S.

Remark 7 It is desirable that all maximal chains have the same length from the genesis to the corresponding tips. Typically some of the tips are not in the same antichain.

The size of the largest antichain is known as the poset's width; The size of the longest chain is known as the poset's height. This will be used in (3).

Remark 8 The height is from the genesis to a tip in the most recent antichain.

A poset can be partitioned using chains (see, e.g., [5–7], etc.). A poset can also be partitioned using antichains which is closely related to our formula construction as shown in [9]. The following theorem is considered self-evident and is presented without the proof.

Theorem 1 (Dual of Dilworth) *Suppose that the largest chain in a poset S has length n. Then S can be partitioned into n antichains.*

Definition 1 A rank function of a poset S is function $r : S \to \{0\} \cup \mathbb{N}$ having the following properties:

(i) if s is minimal, then $r(s) = 0$.
(ii) if t covers s (i.e., $t \gtrdot s$), then $r(t) = r(s) + 1$.

Remark 9 In a DAG, the minimal element is the genesis.

Definition 2 [16] On a finite poset S with length n, ordered by inclusion, the reverse rank function $\rho : S \to \{1, \ldots, n\}$ is defined by

$$\rho(E) = \sum_i \mathbb{1}_{E \subseteq M_i},$$

where E is any element of S and M_i's are incomparable maximal elements of S. The reverse rank function $\rho(E)$, a counting measure, returns the number of maximal elements containing E.

Remark 10 In a DAG, tips are the maximal elements.

Theorem 2 *Given a finite poset $(S, \subseteq)$, $\max_{E \in S} \rho(E)$ is the width (i.e, the size of the largest anti-chain) of a poset and equals to*

$$\min \left\{ m \mid \text{maximal chains } C_1, \ldots, C_m \text{ with } S = \bigcup_{i=1}^m C_i \right\}.$$

Proof Obvious by Dilworth [5] and the reverse rank function in Definition 2.

Definition 3 (*system of distinct representatives*) Suppose that $A_1, A_2, \ldots, A_N$ are sets. The family of sets $A_1, A_2, \ldots, A_N$ has a system of distinct representatives if and only if there exist distinct elements $z^{(1)}, z^{(2)}, \ldots, z^{(N)}$ such that $z^{(i)} \in A_i$ for each $i = 1, \ldots, N$.

Theorem 3 (Duality) *The minimum number of non-redundant edges from an antichain to its previous one is equal to the maximum number of distinct representatives in the latter antichain.*

2.3.2 A Dual Layer Construction and Two Essential Roles

Based on the aforementioned notions, we present the main algorithm which constructs a new layer on top of a convoluted, partially ordered DAG. We will decompose a DAG using chains and antichains. The original DAG is preserved as it normally operates, and there is a dual-layer being constructed almost synchronously based on subset relation among the nodes on the DAG.

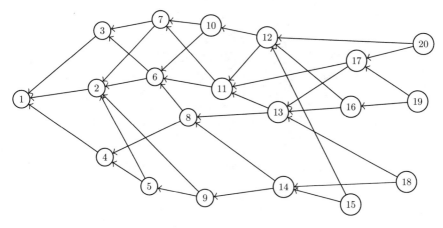

Fig. 4 Transaction units connected in a DAG whose genesis is node 1

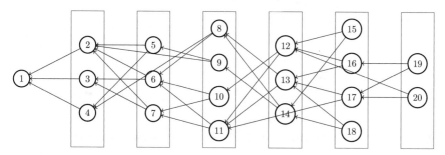

Fig. 5 A series of boxes (antichains) with incomparable units, constructed based on the partial order by their subset relations

Let us consider a DAG of 20 nodes (with four tips of nodes 15, 18, 19, 20) as in Fig. 4. The i-th-created node shows its node number i, i.e, node 1 is the genesis node. Child nodes are the supersets of their parent nodes. Incomparable nodes form an antichain and the set of incomparable child nodes must be a "cover" of the set of incomparable parent nodes. In Fig. 5, for example, the nodes 8, 9, 10, 11 are included in an antichain (the third bluebox from the left) since they are mutually incomparable. The "union" of those four nodes (i.e., antichain) include the "union" of nodes 5, 6, 7 (i.e, antichain)—no single node among the four nodes 8, 9, 10, 11 include all three nodes 5, 6, 7 as a node can only approve two previous nodes.

As described in Fig. 5, by the decomposition using antichains, any entangled DAG can be restructured into a linked list (i.e., a chain of blue boxes in Fig. 5), resembling a blockchain. For this reason, the blockchain is considered as a special version of a DAG, which is a generalized blockchain.

Remark 11 We do not change a given DAG to a single chain; Rather, we preserve the original DAG-based ledger to make the system more secure and trusted. This is the embarkation of our dual ledger-keeping algorithm.

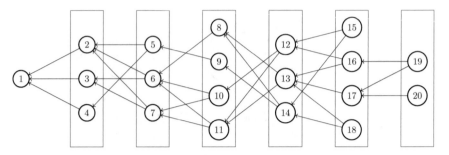

Fig. 6 A series of boxes (antichains) after redundant edges (approvals) are removed

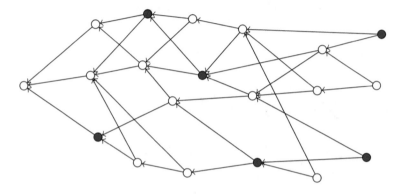

Fig. 7 Primal layer: transaction units in a DAG, our primal space with boxers in blue

Specifically, we present a series of descriptions as in Figs. 4, 5, 6 and 7. Let B_k, $k = 1, \ldots, 6$ denote the bluebox in Fig. 5 from the left to right in order. B_1 has elements of nodes 2, 3, 4, B_2 has elements of nodes 5,6,7, and so on. Then $B_1 \subseteq B_2 \subseteq B_3 \subseteq B_4 \subseteq B_5 \subseteq B_6$ and $B_1 \lessdot B_2 \lessdot B_3 \lessdot B_4 \lessdot B_5 \lessdot B_6$ as a bluebox covers its most recent and previous one.

In including the most recent previous transaction, if the same address issues multiple transactions (multiple nodes) then it is desirable to form a single linked list (i.e., a chain or a totally ordered set). Tracking its own series of transactions will clarify such task.[5]

Figures 5 and 6 are slightly different: A few number of edges in Fig. 5 are missing in Fig. 6. One of the two edges are removed from nodes 8, 9, 17, 20 as some of them are redundant for the sake of the consistency of the system. Note that node 9 approved nodes 5 and 2, but node 5 approved node 2 earlier. Node 9's approval on node 5 means that it automatically approved node 2 as node 5 is the child of node 2. Node 9 is a child of node 5, so node 9 cannot be a child of node 2 since node 2 is

---

[5]A rapid succession of the multiple transactions using different addresses is a typical way to attack the system. More details are discussed in Sect. 2.4.1.

a parent of node 5—it will hurt the parent-child relationship. (The nodes 17, 13, 11 are of the same case.).

The approval selections of nodes 8 and 20 are valid but undesirable. They did not approve redundant transactions and are, thus, harmless in this example. However, in a real situation some idling nodes might select unnecessary and old transactions that are already approved by many other nodes. This may cause a higher complexity in a variety of important operations. Further, this may put nonsensical and illogical cumulative weights on the DAG because it interferes with efficient ordering operation of the system. Selection of nonsensical transactions does not take place in our dual-ledger system, which will be discussed in Sect. 2.4.1.

The blue boxes in Fig. 6 are the antichains as well as the SDRs (if we see distinct chains from the nodes). From the geometric perspective, the elements of the most recent antichain can be considered as vertices of upper bounded orthants in a multi-dimensional space, and it follows that the earlier antichains' elements are inside of such orthants.

Note that there is no link among the elements of the same antichain. Let us pick a single node and appoint it to represent and act for its siblings. The nodes we will choose are the ones that joined their antichains the latest. In this example, such nodes are 4, 7, 11, 14, 18, 20 and they are all over the place in a DAG (See Fig. 7.). We call them the box-closing-members, or "boxers."

Remark 12 (The roles of the box-closing-members, or boxers) The roles of the boxers are adaptable as the system evolves. One of the crucial roles is that they communicate with their siblings, the members of the same antichain, and broadcast to their siblings if there are some notable changes on the system. Moreover, there is a boxers' own network, on which they can efficiently communicate with each other. The network is just a single chain connected by edges among them, but it is a semi-hidden network since it's exclusive for the boxers.

The boxes are the antichains and they can be concatenated to form a chain: the chain of antichains. Individual box-genesis will be denoted by i-genesis (or g_i) if it's inside the i-th box (antichain). The i-genesis is the child of $(i-1)$-genesis and the parent of $i+1$-genesis, and $(i+1)$-genesis is the child of i-genesis and the parent of $(i+2)$-genesis, and so on.

$$\text{genesis} \lessdot \cdots \lessdot (i-1)\text{-genesis} \lessdot i\text{-genesis} \lessdot (i+1)\text{-genesis} \lessdot \cdots \qquad (2)$$

There is also a chain (as in the above covering inequalities in (2)) of box-geneses which is the same height as the whole DAG—The height of a DAG is the size of its longest chain.

Remark 13 (The roles of the box-genesis) The box-genesis (a full-node) is selected among the nodes of good-standing track records. Once selected, one is responsible to check most recently approved transactions by its siblings, followed by announcing the final confirmation of their validity. Such transactions are in the parent-box (most recent previous box). When the final confirmation is placed, the box will finally be closed in agreement with a boxer and no more validations can be placed. More details are presented in Sect. 2.4.1.

If the system has a fixed upper bound of the width or size of antichain, say M, and if the total number of vertices of a given DAG is N, then

$$\text{the height of the whole DAG is at most } \lceil N/M \rceil, \tag{3}$$

which is a markedly important measure owing to the fact that the system governs the quantity M per Monte Carlo simulation. Therefore, the quantity of (3) can only be estimated in the primal space. From the dual space the height of the whole DAG can easily be found, which equals to the number of closed antichains plus two at most. We will use this measure for further discussions, together with another criterion—the time constraint τ for each antichain. This is also a vital constraint to control the system. As both M and τ and their roles are presented in Remark 20, they are both paramount criteria for a boxer selection. The size of antichain will be determined by the box-closing boxer.

This chain of box-geneses is a network from the latest i-genesis (a root node) to the genesis (a sink node and the origin), hidden to users, which helps make the system more secure and compact on the dual layer (in Fig. 8). The regular nodes are not informed of such chain unless the need arises. The detailed roles of the box-geneses are presented in Sect. 2.4.1.

Simply put, the box-making algorithm is as follows: open the box, examine and put items in the right packaging boxes, then close the boxes. Using the boxes based on the algorithm detailed below, we construct a new layer which is a combination of a semi-hidden network and a normal chain. This new layer is a separate space but it corresponds to the original DAG. This layer will be updated in real time as transactions are released and validated on the DAG .

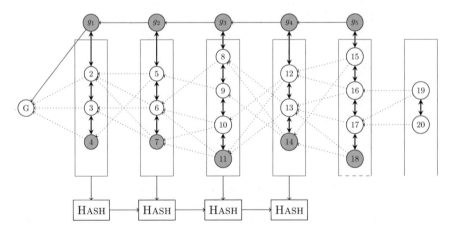

Fig. 8 Dual layer: boxers (in gray) and their siblings in the dual layer, constructed in real time

We present two algorithms—Algorithm 1 at a fixed point of time and Algorithm 2 in real time. The following is to show how to fill a box at a specific time point with unassigned nodes. Let B_i denote the i-th antichain (box) and V_i' the corresponding set of vertices which approved two previous transactions but not yet validated by others. Put another way, V_i' is the set of tips, about to issue new, or unvalidated, transactions.

Algorithm 1 Constructing the i-th box (B_i) at a point of time for $i = 1, \ldots, n$

Suppose the box B_{i-1} is full, and $V_i' = \{v_{i_1}, \ldots, v_{i_n}\}$.
while there exists $v \notin B_i$ for every $v \in V'$ with $B_i \cup \{v\}$ **do**
   if v is a cover of the element(s) of B_{i-1} **then**
      $B_i \leftarrow B_i \cup \{v\}$
      $V_i' \leftarrow V_i' \backslash v$
   else
      $V_{i+1}' \leftarrow V_{i+1}' \cup \{v\}$
   end if
end while
Stop; The box B_i has been filled up, let the last v be the boxer. Then the i-th genesis g_i is assigned, and we move on to the next box B_{i+1} with V_{i+1}'

Remark 14 The box-genesis will be selected among good-standing nodes only after the boxer is determined, which is an imperative rule to keep the system secure. (See Remarks 22 and 25.)

Remark 15 (Timestamp server) Timestamps form a chain, and each timestamp includes the has of previous timestamps as in Fig. 8.

Remark 16 Each box is identified by a hash, and is linked to its previous box by referencing the previous box's hash.

The following is how to build a dual-layer in real time. See Remark 20 for the constraints M and τ for the size of antichains. A node that approves two previous transactions will be assigned to the corresponding antichain (based on the partial order).

Algorithm 2 Constructing the antichains in real time (i.e, a single node at a time)

Suppose $|B_i| = k_i$ for all i.
if a node v just approved two previous transactions, and is a cover of the element(s) of B_{j-1} **then**
   $B_j \leftarrow B_j \cup \{v\}$
   Update $B_j = \{v_j^{(1)}, \ldots, v_j^{(k_j)}, v_j^{(k_j+1)}\}$
end if

Remark 17 (Boxchain on the dual-layer) The dual layer corresponding to the original DAG-based layer has a chain of antichains (boxchain), which resembles a blockchain since each box consists of transactions and such boxes are serially linked.

Similar to Bitcoin transactions, in our dual-ledger system most of the conflicts are easily determined if they are in different antichains (boxes). If they are in different antichains (i.e, totally ordered), the latter one will be rejected. If suspicious transactions are in the same antichain, their cumulative weights can be compared. If such transactions happen to have the same weight (same number of descendants), some tie-break rules can be devised, including a fixed maximum width (the size of the largest antichain), a boxers role, etc. However useful they appear, robust criteria for final confirmation is necessary to build trust among participants and maintain systemic sustainability. For this reason, we develop a dominant consensus mechanism to reach the final stage of agreement which is due to the dual system of both DAG and boxchain ledgers.

Remark 18 (The dual-ledger system) The dual-ledger system consists of a DAG in primal layer and the corresponding boxchain in the dual layer.

2.4 Chain of Antichains

2.4.1 Consensus Protocol for the Final Confirmation with Box-Closing, and a Subtree Problem

As many practitioners and researchers have already pointed out (see, e.g., [3, 25] etc.), there can be a myriad of subtrees, and as a result, a DAG can be extremely wide and inefficient in terms of the final confirmation possibilities. In Fig. 9, there is no path from some latter blue nodes (good-standing nodes) to red nodes (malignant ones). With subtrees, nonsensical transactions can be "issued and approved" by the group of red nodes in the absence of prior investigation of the validity of transactions. This is a major obstacle to reaching a "consensus" on the validity of previous transactions, which is critical in any distributed peer-to-peer system.

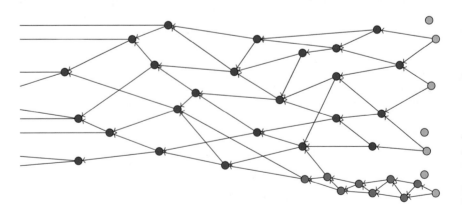

Fig. 9 Main subtree in blue vs. dishonest subtree in red (tips are in gray)

A subtree can occur haphazardly in a DAG, which is prone to unproductive behaviors. Regardless of the intention of forming a subtree, it is imperative to equip a channel to verify the validity of transactions from subtrees. This can be done at the algorithm-design level of the consensus mechanism of reaching decentralized final confirmation of already-approved transactions.

A subtree can take place anytime and a subtree always exists in a DAG, which renders negative behaviors possible. Whether it's formed by a good-will or not (by a good citizen or by a malicious attacker), there must be a way to check transactions' validity from subtrees. Allow us to present what can be done by our particular algorithm regarding a consensus mechanism. This is about reaching a final state of agreement from the peers—the final confirmation of already-approved-transactions.

The main process of our consensus protocol is as follows. For the final confirmation of transactions of box (antichain) B_{i-1}, it is required to reach agreement among the peers in the next box B_i. This means, a final confirmation process for B_{i-1} initiates when the boxer of B_i is selected. To join B_i a node must approve at least one node from B_{i-1}. Thus, the formation of a box is totally based on what transactions the node approved.

Remark 19 (The rule for the tip selection, or transaction selection and validation)) The users are recommended to select most recent transactions. We use the rank function (in Definition 1) for this rule as the following.

$$r(v_k) \leq r(v_{k+1}) \leq r(v_k) + 1, \tag{4}$$

where v_k denotes a node that just finished its transaction validation right before v_{k+1}'s completion of validation. The inequality (4) is equivalent to:

$$v_k \in B_i \longrightarrow v_{k+1} \in B_i \text{ or } B_{i+1}. \tag{5}$$

Once the final confirmation process begins, a later node can no longer select and approve transactions from the corresponding box whose transactions are being checked (Fig. 10).

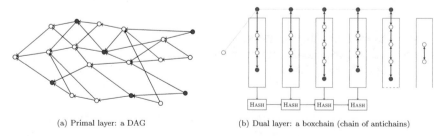

    (a) Primal layer: a DAG   (b) Dual layer: a boxchain (chain of antichains)

Fig. 10 The dual ledger-keeping. Boxers are in blue in both layers; box-geneses are in red in the dual layer

The final confirmation will be placed on each box by its child box (the next neighboring box, or cover). Again, the final confirmation process for B_{i-1} begins when the boxer of B_i is selected. We, thus, list below the following *dual-criteria* for boxer selection.

Remark 20 (Criteria for a boxer selection)

- *Cardinality*: As mentioned in (3), the size of antichains can be determined by the number of nodes. Let M denote the upper limit of the number of elements of the i-th box (B_i). Then we write $|B_i| \le M$. If we only take this into account for a boxer selection, then the Mth node in B_i will be designated as the boxer.

  - For security purpose, M will be selected as follows. Recalling the principle of the inversion method, we compute the value $M = F_X^{-1}(p)$, where F is the cumulative distribution function (c.d.f.) of any discrete random variable X (or a discretized continuous random variable), $l \le X \le u$ where l and u are both positive integers; and p the generated random number around the uniform distribution on $[0, 1]$: It is known that if U is a uniform $(0,1)$ random variable, then with a given c.d.f. F, $F^{-1}(U)$ is a random variable.

- *Duration*: Let the time τ denote the acceptable and agreeable time-limit (e.g., 20 s) for a user to wait until the final confirmation. Note that the average time until the final confirmation would be about 1.5τ. (See Remark 26 for details.) Suppose that the first node v_{i_1} of box B_i comes in at time $t = s$. In other words, at time $t = s$ a node v_{i_1} just finished its validation of two previous transactions in B_{i-1} (or one in B_{i-1} and another one in B_{i-2}). Suppose that a node v_{i_k} finishes the validation of two previous transactions at time $t \le s + \tau$, with the next one $v_{i_{k+1}}$ completing its approvals at time $t > s + \tau$. Then the node v_{i_k} will become the boxer and $v_{i_{k+1}}$ will be the first member of B_{i+1}. In fact the node $v_{i_{k+1}}$ cannot select any transactions from B_{i-1} by (5).

  - No more nodes can be added to the box after whichever criterion is met earlier. The final confirmation process needs to get started when a boxer is selected.
  - **Exception**: If the size M is reached at beyond allowable speed (i.e., in case that incoming transactions are extremely frequent than normal), the time constraint will be used. This is to keep the system safe from possible attacks. (see Sect. 2.4.2).

Note that the size of boxes (antichains) depends on the boxer selection problem as in Remark 20. Transactions issued by the members of B_{i-1} are approved by the members of B_i, but they are just individual validations. We need a strong consensus protocol in order to place the final confirmation of entire set of transactions of the previous box.

Therefore, the whole boxchain (chain of antichains) is marked "true" in a recursive way. Note that there is another recursion inside each box as in the following consensus mechanism.

Algorithm 3 The final confirmation of the transactions in B_{i-1}, conducted by B_i

Step 1. Once the box B_i's size becomes M or time is up ($\geq \tau$), i.e., B_i is full,
the i-genesis (g_i) starts to check its siblings' validations (i.e., validity of transactions of B_{i-1}).

Step 2. If all transactions of B_{i-1} are legitimate based on the consensus protocol (below), go to Step 4. Otherwise, go to Step 3.

Step 3. Disable the nodes with illegal transactions in B_{i-1}, and report it to the genesis group. Disable all nodes in B_i that approved such illegal nodes.

Step 4. The final confirmation is placed by the g_i.
B_{i-1} is closed and marked "true," and every transaction issued in B_{i-1} becomes final, followed by the timestamp synchronization with all the previous box-geneses.

Consensus Protocol in a Dual Recursion

- The 2 + 2 process in recursion:
  When a node approves two previous transactions it will be added to a box as in the Algorithm 2. As the node joins the box, it is automatically augmented to the last node in the box and is required to check the validity of transactions approved by the neighboring node. This process continues recursively from the second to the last node (boxer) of the box. Simply put, each node checks its prior neighbor's approvals upon arrival. This is a dual validation process for every node conducted both in primal and dual layers.
- Final confirmation by the box-genesis:
  The final confirmation of recent transactions will be made by the box-genesis as presented in Algorithm 3. This role resembles that of miners' of Bitcoin, and such validation process is conducted by the box-genesis that only exists in the dual layer. Rewards are given to a box-genesis when the final confirmation is placed.

The final confirmation is placed when the consensus protocol is completed. The box-genesis broadcasts the termination of validation process to all other members. In our peer-to-peer system anyone can take such role (not the miners vs. the rest) as long as one shows a good track-record.

Remark 21 (The box-genesis group membership eligible to good-standing nodes) The roles of box-genesis is crucial to keeping our ecosystem healthy and secure. Excellent history of positive behavior is required for peers to renew this membership and new members with outstanding track-records can also be selected. The box-genesis among the leaders of consensus protocol, and this position is open to everyone but follows a random assignment process.

Remark 22 (Random assignment process) The members of a box-genesis group are not totally trusted even though they qualify—its box assignment is completely random with some conditions, and not known to the box-genesis until the last node of box (boxer) is determined. After the boxer is selected the box genesis will be appointed by a random selection among the nodes other than the boxer.

Remark 23 (Boxer's role for dual safety) The boxer is the one communicating with its siblings (nodes of the box), together with other boxers. However, the final confirmation of validations of previous transactions must be determined solely by a box-genesis. For dual safety, the boxer keeps the final confirmation from the box-genesis in check. Some rewards are given to the boxer.

Remark 24 As we discussed, our recursive consensus mechanism can perform well in terms of both efficiency and scalability. This consensus protocol is unique—in our dual-ledger ecosystem there are no heterogeneous peer groups. Anyone qualified and "lucky" node can play a vital role in the system, e.g., as a box-genesis or as a boxer. It is easy to see that more active peers will have higher chances to get more rewards.

2.4.2 Possible Attack to the System and Its Success Probability

A double-spending attack can cause a serious integrity violation in any distributed peer-to-peer system. Although our recursive validation and final confirmation systems are robust, every possible attack needs to be accounted for in order to keep the network safe. A possible and worst-case attack scenario can be imagined as follows. Suppose that a malicious and affluent party attempts to take over the system over some time-window by issuing an immense number of transactions almost simultaneously with all different addresses. In our dual ledger-keeping system, a complete dominance over a pair of back-to-back boxes (antichains) including the box-geneses and the boxers is necessary for a meaningful damage to transactions.

From the attacker's perspective, the favorable event is that there are no other transactions issued while two consecutive boxes are formed. Suppose that a time constraint τ (in Remark 20) is used for a single box-closing. For simplicity we also assume that the transaction arrivals follow a homogeneous Poisson process. More detailed and realistic, nonhomogeneous compound Poisson processes, etc. are covered in Sect. 3. It is known that the corresponding inter-arrival times are i.i.d. exponential random variables with the same parameter λ of the Poisson process (the average number of transactions in a given unit time). Let a random variable W denote the time until the next transaction, which follows an exponential distribution with a p.d.f. $f(x) = \lambda e^{-\lambda x}$, $x \geq 0$. Then the following can be written:

$$p = P(\text{no transactions while two boxes are formed}) = P(W > 2\tau) = e^{-2\lambda \tau}. \quad (6)$$

Suppose that the system had set a time limit τ of 20 s (1/3 min) for the final confirmation process. Assuming the average number of transactions per minute is 30 (i.e., $\lambda = 30$), we have

$$p = P(W > 2(1/3)) = e^{-2(30)(\frac{1}{3})} \approx 0.000000002061. \quad (7)$$

For a more speedy final confirmation process, let $\tau = 10$ s with the same $\lambda = 30$.

$$p = P(W > 2(1/6)) = e^{-2(30)(\frac{1}{6})} \approx 0.00004539, \quad (8)$$

which is still a sufficiently slim chance. The results clearly demonstrate the following.

- More transactions will lower the probability of a successful attack, i.e., $\lambda \uparrow \Longrightarrow p \downarrow$.
- A longer box-closing interval also brings down the chance of a successful attack, i.e., $\tau \uparrow \Longrightarrow p \downarrow$.

Note that the number of transactions will be sizable in a booming e-commerce marketplace, which makes it enables a very speedy final confirmation process. With this reasoning, let us consider another case, e.g., say 100 transactions per minute on average (i.e., $\lambda = 100$). Then our time constraint τ can be selected so that the probability of the attacker's favorable event is kept negligible. Let us put an upper bound of $p = 0.000001$ (one out of a million) on a successful attack chance. Then we can find τ using

$$P(W > 2\tau) = e^{-2(100)\tau} \le 0.000001, \tag{9}$$

followed by

$$\tau \ge \frac{\ln(0.000001)}{-200} \approx 0.06907755 \text{ min}, \tag{10}$$

which is about 4.14465316 s. This means that if we put on a 5 s rule for the time constraint τ then the attacker's success probability would be less than 0.000001 (again, $\tau \uparrow \Longrightarrow p \downarrow$). If it still seems possible, note that we have a series of box-genesis (two different ones) in the associated back-to-back boxes and they are independent of transactions.

Remark 25 Together with a random selection of the box-genesis, the probability of successful attack would be almost zero in any case.

Detailed stochastic aspects of transaction flows will be discussed in Sect. 3, where one can find more general and useful ideas about how to analyze related random processes.

Remark 26 (Average time until the final confirmation) It is about 1.5τ which is the average of $\tau < T < 2\tau$.

The actual amount of time until the final confirmation for a singe transaction (or a node) is varied depending on the time point when the node joined its corresponding box. Such waiting time until the final confirmation is $\tau < T < 2\tau$ since the final confirmation is made by the next neighboring box (child box). Note that it takes τ to form a single box. It is not hard to see that if a node is the first member of a box, its transaction's final confirmation time would be approximately 2τ since two more boxes need to be closed. In case of a boxer, the confirmation time will be about τ. Recall that the transactions are already approved by their child nodes in a DAG, and therefore our consensus protocol in the dual layer can place a powerful final-confirmation.

Remark 27 In the dual layer, we have a chain of antichains (boxchain) and each antichain (box) consists of a single chain. This means we have a total order in the dual layer and obviously any conflicting nodes can easily be checked.

Let us finalize this section by supporting Remarks 19 and 20 which enable the following.

- No purely random approvals of tips—a lazy user can approve a fixed pair of old transactions
- Inclusion in a more recent antichain, followed by a 2 + 2 validation process.
- In case that a new legitimate transaction is not approved and waits longer than the formation of two new boxes, an empty transaction will be issued as a series. But it still needs to select and approve another node to issue the same transaction in order to get a validation by others. Note that such empty transaction contributes to the network's security.
- M—the width constraint for antichain: This will be determined by the frequency of transaction flows (see Sect. 3 for more details about a random transaction flow).
- τ—the time constraint for antichain: We do not wait until the box is filled up to the given limit M. Recall: If the size M is reached at beyond allowable speed (incoming transactions are extremely frequent than normal), the time constraint had better be used. This is to keep the system safe from the worst case scenario we just studied.

2.4.3 Incentive System with Proper Rewards and Fees

Proper rewards and fees are essential to keep a distributed peer-to-peer network at a desirable status. Note that in Bitcoin "mining" is the incentive mechanism for decentralized security. Our incentive device is for every participant as a mixture of rewards for contribution and fees for usage. The fees in our ecosystem are minimal but serve as an important obstacle to nonsensical orders. Below are the possible cases for rewards. Rewards are given to such user(s) who

- put legitimate validations in the DAG-based primal layer.
- put legitimate validations in the boxchain-synchronizing dual layer.
- complete the job as a box-genesis
- complete the job as a boxer
- report abnormal events.

Recall that the appointment as a boxer or a box-genies is conditionally at random—any qualified user can play these roles multiple times (as mentioned preciously, e.g., see Remarks 12, 13, 21, 22). It is not difficult to see that more active and honest participation will increase the chance to take such positions and receive more rewards, which could end up with more rewards than fees. Fees are paid whenever a transaction is issued. This incentive system is certainly beneficial to both users and the whole ecosystem. Good citizenship is crucial for our purely distributed peer-to-peer system as much as any type of society is in need of.

2.4.4 Our Dual Ledger-Keeping System, Boxchain Compared with Others

The main structure of a blockchain is a single serial link of blocks (hence the name) and each block consists of transactions, therefore the blockchain is a totally ordered set. If we look at each transaction as a single block, there is no longer a single chain; Rather, we see very disorganized, myriads of intertwined blockchains. This generalized blockchain and can be depicted as a DAG, consisting of a set of vertices (transactions) and a set of edges (issue and validation of transactions). The transactions in a DAG are only partially ordered. Indeed, the validation process in a DAG is very efficient and relatively but not as much secure as in the original blockchain. Considered a third generation blockchain after Ethereum as the second, which emerged after blockchain, a number of developers have recently released DAG-based distributed ledgers, associated cryptocurrencies, and applicable payment systems, led by IOTA, Hashgraph, etc.

Irregular and partially ordered DAGs can be reshaped into a more compact form using the discrete mathematical notions of chains, antichains, and SDRs. This idea, coupled with a doubly-secure confirmation protocol, has inspired our dual ledger-keeping system, which is partially based on [14–16]. While preserving the DAG-based ledger, we build another ledger with a chain of antichains to structurally resemble the original blockchain. As a result, along with an effective consensus protocol, our boxchain is an efficient, secure, and decentralized DLT (Table 1).

Table 1 Brief comparison of Blockchain (Bitcoin), IOTA and boxchain

	Blockchain (Bitcoin)	IOTA	Boxchain
Usage	Payment	Used for IoT applications	Payment
Finality	Yes	No	Yes
Confirmation time	10 min at least	2 min (validation)	30 s on average if $\tau = 20$ s (depending on the values of τ and λ)
Transaction fees	0.001 BTC	None	Tiny or none (if rewards are given)
TPS	7	1,000	5,000
Decentralization	Yes	No	Yes
Security	High	Low	High

3   Stochastic Models of the Dual Ledger-Keeping System

3.1   A Nonhomogeneous Poisson Process—Practical Assumptions on the Frequency of Transactions

Unlike some of the assumptions made in [25], we suppose that $X(t)$ is a nonhomogeneous P.P. (Poisson Process), where $\{\lambda(t), t \geq 0\}$ is a stochastic process itself. This is because we believe that transactions' frequency can be varied over certain time periods (e.g., there might be more transactions in the middle of night). For the sake of completeness, allow us to present some basic concepts first. Stochastic process can simply be called "one-parametric family" of random variables $X(t), t \in T$. For example, in case of Markov Chains we have $T = \{0, 1, 2, \dots\}$. In our case we suppose $T = \{t \mid t \geq 0\}$, which means "time" (mathematically it is nothing but nonnegative half of the real line). The following is well known. $X(t), t \geq 0$ is a homogeneous Poisson process if it has two properties: (i) $X(t)$ has independent increments, i.e., for $0 \leq t_0 < t_1 \leq t_2 < t_3 \leq \cdots \leq t_{2n-1} < t_{2n}$, $X(t_1) - X(t_0)$, $X(t_3) - X(t_2)$, $\dots$, $X(t_{2n}) - X(t_{2n-1})$ are independent random variables; (ii) For $s \geq 0, t > 0$, the random variable has Poisson distribution with parameter λt,

$$P(X(s+t) - X(s) = k) = \frac{(\lambda t)^k}{k!} e^{-\lambda t}, k = 0, 1, \dots,$$

which follows that $E(X(s+t) - X(s)) = \lambda t$ and $var(X(s+t) - X(s)) = \lambda t$. Poisson process is a random event process, e.g., transactions occur in the system (our case), customers arrive to a store, cars arrive to a gas station, etc. λ is the expected number of such random events in unit time.

For nonhomogeneous Poisson process, Condition (ii) only needs to be changed. Instead of a constant rate *lambda* we assume that there exists a nonnegative function $\lambda(u), u \geq 0$ such that for $s < t$ $X(t) - X(s)$ has Poisson distribution with parameter

$$\int_s^t \lambda(u)du,$$

which means the parameter depends on time interval. This assumption is crucial in our business because our data set shows some clear trends in frequency of transactions (and their sizes as well). $\lambda(u)$ is called "event density" (in our case, transaction density) because in the interval $(t, t + \Delta t)$

$$\int_t^{t+\Delta t} \lambda(u)du \approx \lambda(t)\Delta t$$

is the expected number of events.

Example 1 (Transaction Density from Data Set) Suppose that the following transaction density (or event density) are found from recent data set (unit time = 1 h).

$$\lambda(u) = \begin{cases} 2u, & 0 \le u \le 1 \\ 2, & 1 \le u \le 2 \\ 4 - u, & 2 \le u \le 4. \end{cases}$$

For simplicity, we are interested only in the time interval [0, 4]. We want to calculate

(i) the probability that two transactions occurred during the first 2 h,
(ii) the probability that two transactions occurred during the second 2 h.

For (i), we need to calculate the event density, say μ, of the first 2 h:

$$\mu = \int_0^2 \lambda(u)du = \int_0^1 \lambda(u)du + \int_1^2 \lambda(u)du = \int_0^1 2u du + \int_1^2 2 du = 1 + 2 = 3.$$

Thus, $P(X(2) = 2) = \frac{3^2}{2!}e^{-3} = 0.2240$. For (ii), we have:

$$P(X(4) - X(2) = 2) = \frac{\left(\int_2^4 \lambda(u)du\right)^2}{2!}e^{-\int_2^4 \lambda(u)du}$$

$$= \frac{\left(\int_2^4 (4 - u)du\right)^2}{2!}e^{-\int_2^4 (4-u)du} = \frac{2^2}{2!}e^{-2} = 0.2707.$$

Remark 28 (Doubly stochastic Poisson processes) A nonhomogeneous Poisson process with the rate function $\{\lambda(t), t \ge 0\}$—it is a stochastic process itself—is called a "doubly stochastic Poisson process." This is a great fit for applications which have "dependent" process increments, e.g., some seasonal products, and new IT products (with business cycles, replaced by newer products), etc.

The simplest doubly stochastic process (sometimes called a mixed Poisson process) has a single random variable θ with $X'(t) = X(\theta t)$, where $\{X(t), t \ge 0\}$ is a Poisson process with $\lambda = 1$. Given θ, X' s a Poisson process of constant rate $\lambda = \theta$, but θ is random (unobservable, typically). If θ is a continuous random variable with p.d.f. $f(\theta)$, the marginal distribution is as follows.

$$P(X'(t) = k) = \int_0^\infty \frac{(\theta t)^k e^{-\theta t}}{k!} f(\theta)d\theta.$$

3.2 Some Important Random Behaviors Regarding Transaction Interarrival Times in a Poisson Process

It is well known that inter-arrival times, let's say $S_0, S_1, \ldots$, are i.i.d. exponential random variables with parameter λ, where λ is the parameter of the Poisson process. Let W_n denote the time of occurrence of the nth event (setting $W_0 = 0$). The differences $S_n = W_{n+1} - W_n$ are the duration that the Poisson process sojourns in state n. See Fig. 11 for description, where $S_0, S_1, \ldots$ denote the inter-arrival times in a Poisson process and $W_1 = S_0, W_2 = S_0 + S_1, \ldots$ represent the occurrence times of the random events in a Poisson process.

We present the following theorems without proofs (see some stochastic models literature for more details).

Theorem 4 *Conditioned on $N(t) = n$, i.e., $N(t) = N((0, t)) = $ number of events in $(0, t)$, the random variables $W_1, \ldots, W_n$ have joint p.d.f.:*

$$g(t_1, \ldots, t_n) = \begin{cases} \dfrac{n!}{t^n} & \text{if } 0 \le t_1 \le \cdots \le t_n \\ 0 & \text{elsewhere} \end{cases}$$

The above theorem tells us that if conditioned on the number of events up to t, i.e., $N(t) = n$, then the n points behave as n independent random points, each chosen from $(0, t)$, according to uniform distribution.

Theorem 5 *Probability density function of sum of n independent, exponentially distributed random variables with the same parameter λ: $X = X_1 + \cdots + X_n$, where $X_i, i = 1, \ldots, n$ have the p.d.f. $f(x) = \lambda e^{-\lambda x}, x \ge 0$. Let $f_n(x)$ be the p.d.f. of X. Then, for $n = 1, 2, \ldots$, we have*

$$f_n(x) = \frac{\lambda^n x^{n-1} e^{-\lambda x}}{(n-1)!}, x \ge 0.$$

It follows that another derivation of probability distribution for interarrival times of transactions. Let $F_n(t) = P(S_0 + \cdots + S_{n-1} \le t)$. Then we have, using Theorem 5,

$$F_n(t) = \int_0^t f_n(u)du = \int_0^t \frac{\lambda^n u^{n-1} e^{-\lambda u}}{(n-1)!}du.$$

Fig. 11 Interarrival times of transactions—exponential random variables in the Poisson process

Furthermore, for $n \geq 1$,

$$
\begin{aligned}
P_n(t) = P(N((0, t)) = n) &= F_n(t) - F_{n+1}(t) \\
&= \int_0^t \frac{\lambda^n u^{n-1} e^{-\lambda u}}{(n-1)!} du - \int_0^t \frac{\lambda^{n+1} u^n e^{-\lambda u}}{n!} du \\
&= \left[\frac{\lambda^n u^n e^{-\lambda u}}{n!} \right]_0^t + \int_0^t \frac{\lambda^{n+1} u^n e^{-\lambda u}}{n!} du - \int_0^t \frac{\lambda^{n+1} u^n e^{-\lambda u}}{n!} du \\
&= \frac{(\lambda t)^n}{n!} e^{-\lambda t}.
\end{aligned}
$$

(11)

Note that $F_k(t)$ means the probability that there are at least k transactions. It follows that $P_n(t) = \dfrac{(\lambda t)^n}{n!} e^{-\lambda t}$ for $n = 1, 2, \ldots$ and this implies $P_0(t) = e^{-\lambda t}$. See our related application below.

Example 2 (Expected total boxcoin in time $(0, t)$) Suppose that transactions occur in the system according to a Poisson process. Upon making a transaction each user pay 1 boxcoin. Find the expected total sum collected in $(0, t)$, discounted back to time 0. Let β be the discount rate. Note that the reason of discounting is to perform more accurate comparison, especially for different time periods.

In case of continuous compounding β is divided by n and then $n \to \infty$. The result is

$$
\lim_{n \to \infty} \left(1 + \frac{\beta}{n} \right)^n = e^{\beta}.
$$

The discounting factor is its reciprocal value $e^{-\beta}$. Discounting from time W_k back to initial time 0, the result is $e^{-\beta W_k}$. We want to calculate the expected total boxcoin in a given time as follows.

$$
\begin{aligned}
E\left(\sum_{k=1}^{X(t)} e^{-\beta W_k} \right) &= \sum_{n=1}^{\infty} E\left(\sum_{k=1}^{X(t)} e^{-\beta W_k} \Big| X(t) = n \right) P(X(t) = n) \\
&= \sum_{n=1}^{\infty} n E(e^{-\beta W_1}) P(X(t) = n) \\
&= \sum_{n=1}^{\infty} n \int_0^t \frac{1}{t} e^{-\beta u} du \, P(X(t) = n) \\
&= \frac{1}{\beta t} (1 - e^{-\beta t}) \sum_{n=1}^{\infty} n P(X(t) = n) \\
&= \frac{\lambda}{\beta} (1 - e^{-\beta t}).
\end{aligned}
$$

(12)

It is well known that if $X_1, \ldots, X_n$ are i.i.d. exponential random variables, then $X_1 + \cdots + X_n \sim \text{Gamma}(n, \mu)$, which has the p.d.f.

$$f(x) = \mu e^{-\mu x} \frac{(\mu x)^{n-1}}{(n-1)!}, \tag{13}$$

where $(n-1)! = \Gamma(n)$ because $\Gamma(n) = \int_0^\infty e^{-y} y^{n-1} dy$ and $\Gamma(n) = (n-1)\Gamma(n-1)$. Its expectation and variance are $E(X) = n/\mu$, $\mathrm{Var}(X) = n/\mu^2$, respectively. Note that the gamma distribution is log-concave, so we can formulate a convex optimization problem using some stochastic optimization techniques.

Gamma distribution, a family member of logconcave distributions.

If $\mu = 1$ of (13), then the distribution is said to be standard. If ξ has gamma distribution, then $n\xi$ has standard gamma distribution. Both the expectation and the variance of a standard gamma distribution are equal to n. An m-variate gamma distribution can be defined in the following way. Let A be the $m \times (2^m - 1)$ matrix the columns of which are all 0–1 component vectors of size m except for the 0 vector. Let $\eta_1, \ldots, \eta_s$, $s = 2^m - 1$ be independent standard gamma distributed random variables and designate by η the vector of these components. Then we say that the random vector

$$\xi = A\eta$$

has an m—variate standard gamma distribution.

3.3 Combining Nonhomogeneous Poisson Processes of Transactions

Multiple transaction inflows into the system must be well analyzed in order to help monitor overall activity of the system. Let us begin with the simplest case and extend to the generality. Let $X_1(t)$ and $X_2(t)$ be the two Poisson flows with parameter λ_1 and λ_2, respectively. We also assume that they are independent. Consider nonoverlapping intervals: (t_0, t_1) and (t_2, t_3). Then $X_1(t_1) - X_1(t_0)$ and $X_1(t_3) - X_1(t_2)$ are independent, and $X_2(t_1) - X_2(t_0)$ and $X_2(t_3) - X_2(t_2)$ are also independent. Since the two processes are also independent, the four random variables are independent. It follows that $X_1(t_1) - X_1(t_0) + X_1(t_3) - X_1(t_2)$ and $X_2(t_1) - X_2(t_0) + X_2(t_3) - X_2(t_2)$ are independent. Thus, we can say that $X(t) = X_1(t) + X_2(t)$ has independent increments. Note that

$$X(s+t) - X(s) = X_1(s+t) - X_1(s) + X_2(s+t) - X_2(s),$$

where $X_i(s+t) - X_i(s)$ has Poisson parameter $\lambda_i t$, $i = 1, 2$. Hence $X(s+t) - X(s)$ has Poisson distribution with parameter $(\lambda_1 + \lambda_2)t$.

The procedure is the same in the general case. We can unite arbitrary finite number of independent Poisson process: if $X_1(t), \ldots, X_n(t)$ are independent Poisson processes with parameters $\lambda_1, \ldots, \lambda_n$, respectively, then $X(t) = X_1(t) + \cdots + X_n(t)$ is a Poisson process with parameter $\lambda_1 + \cdots + \lambda_n$.

3.4 Transaction "size", and the Recursive Formulas for the Probability Mass Function of Compound Random Variables

Transactions are randomly occurring events. Since a number of events occurring in a fixed period of time, say N, is uncertain, it would be suitable to use a compound distribution modelling for a random sum $S = X_1 + X_2 + \cdots + X_N$, where N is a nonnegative integer-valued random variable. For the frequency of transactions, let $N(t)$ be a Poisson process with a fixed rate λ, for simplicity. It is not a problem, however, if the Poisson parameter λ is also uncertain as discussed in the previous sections.

If we need to consider the magnitude (or size) of transactions but their sizes are random (or unknown), it is suitable to use a compound Poisson process. The following is well-known [10]:

$$N \sim \text{Poisson}(\lambda), \text{ where } \lambda \sim \text{Gamma}(r, p/(1-p)) \rightarrow N \sim \text{NegBinomial}(r, p),$$

which means that if the Poisson parameter λ is uncertain due to the behavior of heterogeneous users, a gamma distribution may be suitable for capturing the λ information. Then N will have a negative binomial distribution. Negative binomial distributions can be used for our model with no problem, but we restrict ourselves to the compound Poisson distribution for this paper.

The frequency of transactions plays a central role and in our view, the size of transactions would also be very important to proper management of a DAG. For the transaction size, we let $X_1, X_2, \ldots$ denote independent, identically distributed random variables (meaning the transaction sizes at the corresponding events), which are also independent of the Poisson process. (i.e., the random variables $N, X_1, X_2, \ldots$ are mutually independent.) Then $S(t) = \sum_{k=1}^{N(t)} X_k, \forall t \geq 0$ (i.e., $S(t)$ is a compound random variable) means the aggregate transaction weight in time interval $(0, t)$. It is well known that $E(S(t)) = \lambda t \mu$, $var(S(t)) = \lambda t(\sigma^2 + \mu^2)$ if $\mu = E(X_k), \sigma^2 = var(X_k), k = 1, 2, \ldots$.

In order to calculate the distribution of aggregate transaction weights we use Panjer's recursion formula (see for details, e.g., [2, 4, 20, 22], etc.). For completeness, allow us to present here some basic notions and related formulas. The size X_k can be fitted with continuous or discrete distributions, but the continuous distribution needs to be discretized for the use of Panjer's recursion formula. This is a great fit for our model because every transaction has a positive integer (its weight). For practical discretization methods we refer the reader to the literature, e.g., [4, 21]. We restrict our attention to the case that the transaction size X_k follows a discrete distribution on the positive integers because a continuous variable needs to be discretized to use the recursion formula and also because any positive integer-valued variables can easily be scaled to the suitable size. Then S is also distributed on the nonnegative integers, and the probability mass function of the compound process $S(t)$ can be calculated recursively by the following well-known recursion:

$$P(S(t) = k) = \frac{1}{k}\lambda \sum_{i=1}^{k} i P(X_1 = i) P(S(t) = k - i). \tag{14}$$

Given a Poisson process $N_j(t)$ and nonnegative integer-valued random transaction size X_{jk}'s for the upcoming time periods j, $j = 1, \ldots, M$, we can write:

$$P(N_j(t) = x) = \frac{(\lambda_j t)^x}{x!} e^{-\lambda_j t}, \ x = 0, 1, \ldots; \ j = 1, \ldots, M$$

$$S_j(t) = X_{j1} + X_{j2} + \cdots + X_{jN_j(t)}, \ j = 1, \ldots, M, \tag{15}$$

where X_{jk} is the kth transaction size in the jth process. See Fig. 12 for description.

Let $f_j(x) = P(S_j(t) = x)$, where x is a positive integer. Then, the recursion formula for the p.m.f $f_j(x)$, $j = 1, \ldots, M$ can be written as:

$$f_j(x) = \frac{\lambda_j}{x} \sum_{k=1}^{x} k p_j(k) f(x - k), \ x = 1, 2, \ldots,$$

$$f_j(0) = e^{-\lambda_j}, \ j = 1, \ldots, M, \tag{16}$$

where $p_j(k) = P(X_{j1} = k)$.

4 Models and Discussions for Boxdollar

4.1 Background of Boxdollar

Boxdollar, a value-preserving medium of payment, store of value, or stablecoin, will be used primarily as a useful medium of exchange as well as a dependable store

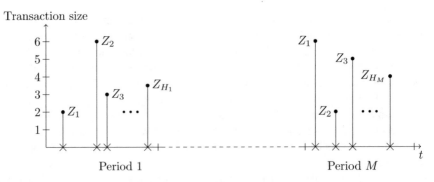

Fig. 12 Illustration of compound Poisson distributed transactions. H_j is the number of events incurred over the period j and the unit transaction size is one boxcoin

of value. The value of boxdollar is at a fixed exchange rate one to one to a fiat currency, hence boxdollar is a fiat money-backed asset and keeps its stability. The goal of analytical models of boxdollar is twofold currency conversion and currency rebalancing problems. These models are well known, and therefore we list up a few models in this section. A variety of useful optimization models for financial applications, which can be found in the operations research literature (see, e.g., [24, 26, 29], etc.).

P2P payments are among impactful applications due to the similarity of operations and transactions on a decentralized system structure: They have chains of digital signatures of asset transfers, P2P networking, etc. Yet, most of P2P platforms are not fully utilizing the "purely" distributed system. There are numerous of P2P platforms which made it easier and faster to make payments, send or receive funds. They provide us with clearly better solutions than traditional systems in most aspects, even with lower fees and costs. Still, their pricing and operating models could be more efficient and secure by the blockchain technology, which may lead us to find a more perfect solution.

Most of existing cryptocurrencies are not accepted in regular market places due to their high volatility. Holding some cryptocurrency is extremely risky, so the current forms of cryptocurrencies are not suitable to use in our daily life. It follows, therefore, the price stabilization is the key to the application of a cryptocurrency. A cryptocurrency, with guaranteed price stability, will clearly be beneficial. The electronic commerce (or e-commerce) industry is where one can find it suitable to apply the blockchain technology and cryptocurrency to its existing P2P payment system. Most of successful e-commerce marketplaces have flexible and adaptable structure, and more importantly, with less regulation but higher potential. For this reason, the e-commerce is a great t for incorporating the ideas of cryptocurrency and its related applications. At this moment the e-commerce market transactions are mostly based on at currencies, while there has been a serious need of instantaneous transactions and borderless transfer-of-ownership. Yet, there is a variety of difficulties in the international transfer, which often challenges tracking, and normally requires more time (than domestic) with higher cost to exchange, various fees depending on a sending or receiving financial institution.

For instantaneous transactions, possible real-life applications include supply chain in drop-shipping, manufacturing management, insurance claim payment process, tax collection processes, and many others. The existing models are not well-suited for our daily use on such applications, and this inspired us of a new stablecoin.

4.2 Currency Conversion Problem

We may encounter situations of needing to exchange boxdollar to some other fiat currency (possibly to multiple currencies)—observation of nontrivial movement in exchange rates, our market participants' requests, and/or regular portfolio (multicurrency accounts) rebalancing, etc. When it comes to currency exchange, the loss

Fig. 13 Stable boxdollar
matching with various fiat
currencies: Chinese yuan
(red), Japanese yen
(yellow), Korean won (blue), Hong
Kong dollar (green), Taiwan
dollar (purple)

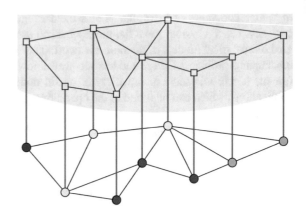

of stabilization is what we're most concerned about. This means we should always maximize the US dollar value (i.e., the amount of boxdollar) of desired positions in converting to other currencies. It is very important that we obtain the best currency conversion strategy to keep up with our stable boxdollar models. Such currency conversion problems (with possibility of arbitrage detection) are well known in operations research literature (Fig. 13).

In what follows, we present optimization problems suitable to rebalance the combination of multinational currencies. Their optimal solutions (i.e., optimal weights on currencies) will be employed for a conversion problem. Note that currency rebalancing will take place only if there is a guaranteed capital gain.

4.3 Currency Rebalancing—Kataoka's Problem: The Safety First Model

Typically random variables appear only on the right hand sides of constraints of stochastic program. The following problem was formulated by [12] and it has stochastic constraints where their technology matrix has random variables. Many real life applications can be solved, but we restrict ourselves to our problem in this paper. Let x_i denote the weight of the ith currency, $i = 1, \ldots, n$ and the random vector ξ consists of components meaning the return on holding the corresponding currency. Our model is the following:

$$
\begin{aligned}
&\max d \\
&\text{subject to} \\
&P\left(\sum_{i=1}^{n} \xi_i x_i \geq d\right) \geq p \\
&\sum_{i=1}^{n} x_i = 1, x \geq 0,
\end{aligned}
\tag{17}
$$

where we assume that $\xi = (\xi_1, \ldots, \xi_n)^T$ has an n-variate normal distribution with

$$\mu_i = E(\xi_i), \ i = 1, \ldots, n, \mu = (\mu_1, \ldots, \mu_n)^T,$$

$$C = E(\xi - \mu)(\xi - \mu)^T.$$

Note that p and M are constants and the decision variables are $x_1, \ldots, x_n, d$. Using some mathematical steps (see, e.g., [27]) the formulation (17) can be written up as

$$\max \left\{ \mu^T x + \Phi^{-1}(1 - p)\sqrt{x^T C x} \right\}$$
$$\text{subject to} \tag{18}$$
$$\sum_{i=1}^{n} x_i = 1, x \geq 0.$$

Since C is a positive semidefinite matrix, the function $\sqrt{x^T C x}$ is convex. When the probability level is set $p \geq 0.5$ (i.e., $\Phi^{-1}(1 - p) \leq 0$), the objective function is concave so (18) turns out to be a convex programming problem.

4.4 Currency Rebalancing—Conditional Value-at-Risk, I.e., Minimization of Risk

Optimization problems using Conditional Value-at-Risk (CVaR) have been researched and used in practice, so allow us to present the formulation of CVaR. We recommend the readers to the literature, e.g., [23, 28, 31] and the references therein. In order to find the optimal portfolio using the classical CVaR, let ξ denote the loss vector and x the currency weights (the decision vector) for a portfolio of n currencies. Then $\xi^T x$ means the loss of the currency holding, and the following model can be written:

$$\min_{x,a} \left(a + \frac{1}{1 - p} E \left([\xi^T x - a]_+ \right) \right)$$
$$\text{subject to } \mu^T x \leq \mu_0$$
$$\sum_{i=1}^{n} x_i \leq 1 \tag{19}$$
$$x \geq 0,$$

where $\mu = E(\xi)$ and μ_0 is some constant. Note that $a = \text{VaR}_p(X)$ at optimality and the optimal objective value is the smallest among all values of $E(\xi^T x \mid \xi^T x \geq \text{VaR}_p(\xi^T x))$ with x in the feasible set of the constraints. Also note that we can write $\mu_0 = -R$ where R means the minimum required return for the portfolio.

It is well known that the following LP is a discrete version of (19) with $\mu_0 = -R$,

$$\min a + \frac{1}{K(1-p)} \sum_{k=1}^{K} u_k$$

$$\text{subject to } a - x^T \mathbf{y_k} + u_k \geq 0, \ k = 1, \ldots, K$$

$$u_k \geq 0, \ k = 1, \ldots, K \tag{20}$$

$$-x^T \mu \geq R$$

$$\sum_{i=1}^{n} x_i \leq 1$$

$$x_i \geq 0, i = 1, \ldots, n,$$

where $\{\mathbf{y_1}, \ldots, \mathbf{y_K}\}$ denotes K i.i.d. samples of the loss random vector $\xi \in R^n$. (20) can equivalently be written in the following matrix notation:

$$\min a + \frac{1}{K(1-p)} \sum_{k=1}^{K} u_k$$

subject to

$$\begin{pmatrix} 1_K & -\mathbf{Y} & I_K \\ 0 & -\mu^T & 0 \ldots 0 \\ 0 & -1_n^T & 0 \ldots 0 \end{pmatrix} \begin{pmatrix} a \\ x \\ u \end{pmatrix} \geq \begin{pmatrix} 0 \\ R \\ -1 \end{pmatrix} \tag{21}$$

$$u \geq 0, x \geq 0,$$

where I_K means $K \times K$ identity matrix, Y is a $K \times n$ matrix with samples of the loss random vector $\xi \in R^n$, and 1_K is a $K \times 1$ all ones vector.

4.5 More General Formulations

Let x denote the vector of currency weights with its related cost vector c, and ξ the random vector with an estimated distribution; the matrices A and T are present and future constraints, respectively. Then a more general stochastic programming model is formulated in the following way:

$$\min c^T x$$

$$\text{subject to } Ax = b, x \geq 0 \tag{22}$$

$$P(Tx \geq \xi) \geq p,$$

where p is a fixed probability chosen by ourselves. In practice p is near 1, for example we may choose p as 0.8, 0.9, 0.95, 0.99, depending on our reliability requirement, i.e., in what proportion of the cases do we want the inequality $Tx \geq \xi$ to be satisfied. Let T_i denote the ith row of matrix T and ξ_i the ith component of random vector ξ.

There is a simplification possibility for the problem (22). Instead of $P(Tx \geq \xi) \geq p$ we take

$$E(\xi_i - T_i x \mid \xi_i - T_i x > 0) \leq d_i, i = 1, \ldots, r.$$

If the function $g_i(z) = E(\xi_i - z \mid \xi_i - z > 0)$ is decreasing, then $g_i(T_i x) = E(\xi_i - T_i x \mid \xi_i - T_i x > 0) \leq d_i$ is equivalent to $T_i x \geq g_i^{-1}(d_i), i = 1, \ldots, r$ and the whole problem becomes:

$$\begin{aligned} &\min c^T x \\ &\text{subject to } Ax = b, x \geq 0 \\ &\qquad T_i x \geq g_i^{-1}(d_i), i = 1, \ldots, r, \end{aligned} \tag{23}$$

which is an LP.

Remark 29 *(Related measures of violation)* In reliability theory and insurance problems $E(\xi - t \mid \xi - t > 0)$ is called "Expected Residual Lifetime." It is natural that it is a decreasing function of t, but it is not always decreasing. (i.e., there are probability distributions for which it is not true, e.g., lognormal, Pareto if $t \geq 1$, etc.) If ξ has a **logconcave p.d.f.**, then $E(\xi - t \mid \xi - t > 0)$ is a decreasing function of t.

Note that the problem (22) is a special case of the following general formulation:

$$\begin{aligned} &\min h(x) \\ &\text{subject to} \\ &h_0(x) = P(g_1(x, \xi) \geq 0, \ldots, g_r(x, \xi) \geq 0) \geq p_0 \\ &h_1(x) \geq p_1, \ldots, h_m(x) \geq p_m, \end{aligned} \tag{24}$$

where x is the decision vector, ξ is a random vector, $h(x), h_1(x), \ldots, h_m(x)$ are given functions, $0 < p_0 \leq 1, p_1, \ldots, p_m$ are given numbers.

Remark 30 (Convexity of the problem (24)) Any logconcave function is quasi-concave, hence if $\xi \in R^r$ has a continuous distribution and logconcave density then $h_0(x)$ in problem (24) is quasi-concave. Hence, $h_0(x)$ allows for the convex programming property. If the objective function h is convex and we assume that $h_1, \ldots, h_m$ are quasi-concave, then the problem is indeed convex.

Remark 31 If the random vector ξ has independent components $\xi_1, \ldots, \xi_r$, then

$$P(Tx \geq \xi) = \prod_{i=1}^{r} P(T_i x \geq \xi_i) = \prod_{i=1}^{r} F_i(T_i x), \tag{25}$$

where F_i is the c.d.f. of ξ_i for $i = 1, \ldots, r$. The probabilistic constraint takes the form:

$$\prod_{i=1}^{r} F_i(T_i x) \geq p. \tag{26}$$

x is feasible if this inequality, in addition to $Ax = b, x \geq 0$, is satisfied.

Note that a variety of optimization applications can be found in [26] and the author's numerous excellent scientific articles.

4.6 Valuation of Boxdollar

The value of boxdollar ($\mathcal{B}_0$) is initially pegged against a fiat currency ($\mathcal{M}_0$) held in the custodian vault. The deposited collateral will generate interest (r). Thus, according to the cost-of-carry theory, an arbitrage-free condition entails transfer of interest to the users of boxdollar, otherwise boxdollar will be discounted as much as to compensate for opportunity costs as follows:

$$\mathcal{B}_0 = \mathcal{M}_0 \implies \mathcal{B}_t = e^{rt}\mathcal{M}_0. \tag{27}$$

However, as the users of boxdollar benefit from its convenience, the yield of convenience (δ) can partially or utterly defray foregone interest rate or opportunity costs. In that case, the value of boxdollar in the future ($\mathcal{B}_t$) can maintain the initially pegged amount ($\mathcal{M}_0$) as such:

$$\mathcal{B}_0 = \mathcal{M}_0 \implies \mathcal{B}_t = e^{(r-\delta)t}\mathcal{M}_0 \approx \mathcal{M}_0 \text{ if } \delta \approx r. \tag{28}$$

4.7 Boxdollar as a Regional Currency and a Quasi-fiat Money

Boxdollar can be used as a regional currency with collateral on tax revenues. This way, the regional government can wield a multiplier effect for increased transactions of goods and services in the community. The key to succeeding a liquid, well-trusted local money lies in the trust among community members and the eventual guarantee of conversion to the fiat currency by the local government. In a well-coordinated game-theoretic setting the agreement among economic agents can create value, as the local currency can boost the regional economy as an "outside" money [8]. A municipality with stable tax revenues can be an ideal place to adopt a cryptocurrency as its local currency. For example, Ithaca, NY—where "Ithaca hours" is circulated as a local money for shopping in local stores—can consider converting the conventional, note-based hours into a cryptocurrency for wider and frictionless usages. Those local governments with liquid, local cryptocurrencies can later enter into mutual agreements to render their currencies compatible. As mutual trust grows and strengthens, the locally and inter-municipally used cryptocurrencies can become a quasi-fiat money across the nation.

Suspicions as per whether a cryptocurrency can become a fiat money is not a new question as the history of finance has witnessed analogous events throughout the time. DLTs utilizing computers scattered around the globe to revise the same list of transactions appeal to finance's fundamental objective of ascertaining a transaction between two mutually-remote parties trustworthy to each other. The predecessors of DLTs have done similarly as banks in England established a clearinghouse in the late 18th Century to revamp the system of managing different ledgers of common transactions [13]. These banks issued new bank notes ("wagon-way through the air") based on the collateral of gold coins ("earth-bound highway", [30]). In another latter century, those who wanted to supply more money ("currency school") clashed with their counterpart whose policy goal was to put the gushing bank notes on hold ("banking school"), which was a precursor of "forking" on cryptocurrencies.

Through trials and errors, inflation control as the foremost policy target of central banking has confirmed a general equilibrium-theoretic prediction of the role of money as a numéraire of the modern economy such that goods and services can be valued at their absolute prices, not the relative prices or exchange ratios of a batter economy. Governance is the key to maintaining the banking system based on a currency. For a fiat money, it is the role of the central bank to grant undoubting trust on the money as a medium of payment and exchange, and a store of value. For a cryptocurrency, the governance of its ecosystem is the protocol of transaction confirmation. For our dual ledger-keeping boxchain, its "2 + 2" doubly-secure consensus protocol is the crux of governance or security which emboldens trust among the participants across the system of utility and incentives.

4.8 The Dual Cryptocurrencies: Boxcoin and Boxdollar in the Value Chain

Boxdollar, a stablecoin will be used primarily as a medium of exchange. The participants will exchange their local fiat currencies to boxdollar based on a real time conversion rate to USD. This is quite simple but strong enough to obtain desirable level of trust and security—one can put down a deposit with a US Dollar for every boxdollar issued (i.e., 1 to 1) so that the boxdollar is asset backed and keeps its stability. There will be a unique and efficient digital wallet, called the boxpay-wallet. Using the boxpay-wallet, the market participants will see all previous transactions, conversion records, and current balances of boxdollar and boxcoin. If registered, fiat currencies in his or her bank accounts can be seen as well. The real time exchange rates of boxdollar to boxcoin, boxcoin to boxdollar and ones among all related fiat currencies are also presented on the boxpay-wallet.

The boxpay protocol consists of the most efficient dual currencies—boxcoin and boxdollar on public and private networks, respectively. In what follows we list up some of the notable functions of boxdollar and boxcoin. Boxdollar has the following desirable features:

- Reliability and security
- Medium of exchange
- Store of value

- Diversification of currency holdings
- Transferability (e.g., efficient transfer (cross-border) in P2P transactions, among merchants and customers, etc.)
- Tracking all previous transactions.

Boxcoin is mainly for making micropayment on public network (see Sect. 2 for details). There are some specifics for the use of boxcoin, the key functions and benefits of boxcoin include the following:

- Rewards and fees (as presented in Sect. 2.4.3)
- Possibility of capital appreciation.

Let us see from a standpoint of a buyer, say Alice. After she exchanged her local currency to boxdollar (BXD), based on the conversion rate to USD 1, the calculated amount of BXDs will be stored in her boxpay-wallet and be ready to use. If Alice wants to use her local currency Chinese yuan (CNY) to buy an item priced at BXD 500, first thing to do is to spend CNY 3,440 to receive BXD 500 into her boxpay-wallet. (Buying 1 US dollar for Chinese yuan requires CNY 6.88 using the conversion rates as of 1 PM (UTC -4), 8/14/18.) Note that it might be a good idea to put down more CNY to get more BXD if the USD appreciation is expected. (For example, the rate USD 1 = CNY 6.88 at the moment may later be changed to 1 USD > CNY 6.88. It's been a while the USD gets more valuable compared to the other currencies.) She may have multiple items in her to-buy list even if she does not want to buy them at the moment. After transaction completed the seller will be able to see the BXDs from the boxpay-wallet, ready to use in the e-marketplace (e.g., for shipping cost) or get an exchange for some local fiat currency (or multiple currencies) as needed.

Remark 32 (The boxpay wallet and smart insurance) The boxpay wallet is now being developed, and is a file (a simple database) of the digital keys, which are completely independent of the protocol. This will come with a smart insurance capability, systematically identifying claims to report.

5 Concluding Remarks

We are in transition to a cashless society and, for quite some time, blockchain has been around as one of the most thriving technologies with a bright future ahead. Blockchain is a decentralized-distributed system which has revolutionized our perspective of the world. Many forms of digital currency have already been used in a variety of ways and places, e.g., in the online marketplaces and mobile banking systems, where it is now normal to use a phone number or an email address instead of a bank account. Although there are numerous benefits the digital currency has to offer, many cryptocurrencies have very low trading volumes as they failed to build their own ecosystem.

In this paper we introduce our new way of thinking on approach to DLT—the dual ledger-keeping algorithm of the chain of antichains (boxchain). Our DLT uses the original transactions occurred in the DAG-based primal space. Using the dual ledger-keeping algorithm a chain of antichains is constructed in real time in the dual layer that synchronously reflects transactions in the DAG. This chain of antichains has resembles the structure of blockchain. Using such dual-layer framework, we take desirable aspects from both blockchain and a DAG-based distributed network system. We consequently arrive at a powerful consensus protocol which makes the final confirmation feasible with great efficiency. Our dual approach proposes two distinct cryptocurrencies: boxdollar and boxcoin. Our stablecoin boxdollar is pegged to a fiat money, anticipating the use as a medium-of-exchange as well as a reliable store-of-value. Another cryptocurrency, called boxcoin, is a crucial component to keep up a purely distributed peer-to-peer network, with capability to run our unique incentive system at its core, indispensable to an effective DLT.

We presented both deterministic and stochastic aspects of the dual ledger-keeping. Illustrative and numerical examples are also presented. Our new algorithms were discovered by the use of discrete mathematics as well as probability models. This paper is focused on the mathematical and analytical foundations of a new digital ecosystem. We hope our new ideas for DLT will be beneficial to the readers and be helpful to improve decentralized-distributed network systems and their real-life applications.

Acknowledgements The first and corresponding authors appreciate research support from the LeBow College of Business, Drexel University, and the College of Business, Ewha Womans University, respectively.

References

1. Baird L (2016) Hashgraph consensus: fair, fast, byzantine fault tolerance. Swirlds tech report TR-2016-01
2. Bowers NL Jr, Gerber HU, Hickman JC, Jones DA, Nesbitt CJ (1997) Actuarial Mathematics. The society of Actuaries, Schaumburg, Illinois
3. Churyumov A (2017) Byteball: a decentralized system for storage and transfer of value. http://byteball.org
4. Dickson DCM (1995) A review of Panjer's recursion formula and its applications. Brit Act J 1(1):107–124
5. Dilworth RP (1950) A decomposition theorem for partially ordered sets. Ann Math 51(1):161–166
6. Dilworth RP (1960) Some combinatorial problems on partially ordered sets. Proc AMS Sympos Appl Math 10:85–90
7. Fulkerson DR (1956) A note on Dilworth's theorem for partially ordered sets. Proc Amer Math Soc 7:701
8. Goodhart CAE (1989) Money, information and uncertainty, 2nd edn. Macmillan, London
9. Greene C, Kleitman D (1976) The structure of Sperner k-family. J Comb Theory Ser A 20:80–88
10. Greenwood M, Yule GU (1920) An inquiry into the nature of frequency distributions representative of multiple happenings with particular reference to the occurrence of multiple attacks of disease or of repeated accidents. J R Stat Soc 83(2):255–279
11. Iota (2016) A crytocurrency for internet-of-things. http://www.iotatoken.com
12. Kataoka S (1963) A stochastic programming model. Econometrica 31:181–196
13. Kindleberger CP (2015) A financial history of Western Europe. Routledge

14. Lee J (2017) Computing the probability of union in the n-dimensional Euclidean space for application of the multivariate quantile: p-level efficient points. Oper Res Lett 45:242–247
15. Lee J, Kim J, Prékopa A (2017) Extreme value estimation for a function of a random sample using binomial moment scheme and Boolean functions of events. Discrete Appl Math 219(11):210–218
16. Lee J, Prékopa A (2017) On the probability of union in the n-space. Oper Res Lett 45:19–24
17. Lerner SD (2015) Dagcoin: a crytocurrency without blocks. http://bitslog.wordpress.com/2015/09/11/dagcoin
18. Nakamoto S (2008) Bitcoin: a peer-to-peer electronic cash system. http://www.bitocin.org
19. NXTFORUM.ORG, PEOPLE ON (2014) Dag, a generalized blockchain. http://nxtforum.org/proof-of-stake-algorithm/dag-a-generalized-blockchain/
20. Panjer H (1981) Recursive evaluation of a family of compound distributions. AST IN Bulletin 12:21–26
21. Panjer H, Lutek B (1983) Practical aspects of stop-loss calculations. Insur Math Econ 1:159–177
22. Panjer H, Willmot GE (1986) Computational aspects of recursive evaluation of compound distributions. Insur Math Econ 5 113–116
23. Pflug GCH (2000) Some remarks on the value-at-risk and conditional value at risk. Prob Constrained Optim 272–281
24. Pflug GCH (1996) Optimization of stochastic models—the interface between simulation and optimization. Kluwer Academic Publishers
25. Popov S (2017) The tangle. http://www.iotatoken.com
26. Prékopa A (1995) Stochastic programming. Kluwer Academic Publishers
27. Prékopa A (2003) Probabilistic programming. In: Ruszczyński A, Shapiro A (eds) Hand books in operations research and management science, vol 10, pp 267–351
28. Prékopa A, Lee J (2018) Risk tomography. Eur J Oper Res 265(1):149–168
29. Shapiro A, Dentcheva D, Ruszczyński A (2009) Lectures on stochastic programming: modeling and theory. SIAM and MPS, Philadelphia
30. Smith A (1776) An inquiry into the wealth of nations. W. Strahan and T. Cadell, London
31. Uryasev S, Rockafellar RT (2000) Optimization of conditional value at risk. J Risk 2:21–41

Blockchain for Intelligent Gas Monitoring in Smart City Scenario

Ashutosh Mishra, Rakesh Shrestha, Shiho Kim and Navin Singh Rajput

Abstract The increasing urbanization demands development of smart cities. Smart cities can be considered as to serve the requirement of its citizen in better way. The smart cities have many applications involving intelligent gas monitoring. In this chapter we will find about the intelligent gas monitoring in smart city scenario. We will see the aspects of smart gas monitoring, understand the concept of gas sensing, requirement of gas monitoring viz., classification and quantification of gases/odors and the brief introduction about the gas sensing. Further, we will understand the application of blockchain in the intelligent gas monitoring.

Keywords Blockchain · Gas sensing · Intelligent gas monitoring · Smart city

1 Introduction

The increase in population and the rapid transition to urbanization requires the smart planning, and development of cities. The smart cities are therefore, the concept of planning and development of cities to serve the citizen in smarter way [1]. This idea of smart city stimulates several responsible upon the professionals (like, architects and designers, engineers, environmental scientists, government, etc.) [2]. Researchers have plenty of unsolved problems to tackle to fulfill the smart city requirements. In this chapter we have focused on Gas monitoring in smart cities. It plays a very essential role in the smart city scenario because of its multifaceted applications. It

A. Mishra (✉) · R. Shrestha · S. Kim
School of Integrated Technology, Yonsei Institute of Convergence Technology, Yonsei University, Incheon, South Korea
e-mail: ashutoshmishra@yonsei.ac.kr

R. Shrestha
e-mail: rakez_shre@yonsei.ac.kr

S. Kim
e-mail: shiho@yonsei.ac.kr

N. S. Rajput
Department of Electronics Engineering, Indian Institute of Technology (BHU), Varanasi, India
e-mail: nsrajput.ece@iitbhu.ac.in

© Springer Nature Singapore Pte Ltd. 2020
D. Singh and N. S. Rajput (eds.), *Blockchain Technology for Smart Cities*, Blockchain Technologies, https://doi.org/10.1007/978-981-15-2205-5_3

has several important applications and we will see them in detail within this chapter. Also, we will discuss about the intelligent gas monitoring and the related security issues here in this chapter.

1.1 Smart City

The concept of smart cities started from the late 90s. The increasing demands of urbanization desire many applications where smartness can be introduced to make better utilization of the available resources. The concept of smart city basically incorporates information and communication technologies (ICT) in such a manner to enhance the quality and performance of daily urban services. For example, the daily energy requirements, transportation, and other utilities etc. The optimized utilization of these services in order to reduce resource consumption, wastage and overall costs leads the basic concept of smartness. Therefore, aim of the smart city is to enhance the quality of living of its citizens through smart technology [1, 2]. Actually, because of the daily development in the technologies there is no exact explanation of 'smart city'. Smart cities employ technologies to improve the quality transport, traffic management, environment, etc. Now-a-days the smart city has many names such as Digital City, Information City, Intelligent City, Knowledge-based City, Ubiquitous City, Wired City etc. According to Mark Deakin, "A city that utilizes ICT to meet the demands of its citizens, and that community involvement in the processes is a necessity for a smart city" [3]. He suggested that following factors can be considered to classify a city as smart city:

- Application of digital and electronic technologies in the services of the city
- Involvement of ICT for enhancing the citizen lifestyle
- Smart working environments for people in the city
- Updated and ICT supported government systems.

Therefore, the daily development of technologies and scientific research provide us a very good plethora to fulfill the requirements. Also, with the increasing interdisciplinary researches many of the solutions are possible to meet the smart city scenarios. Technologies like 5G, Internet of Things (IoT), Artificial Intelligence (AI) supports the smart things. Therefore, the smart cities have smart metering system, autonomous vehicular transportation, smart grid communications, real-time location and utilization, smart parking, smart traffic managements etc. Further, air and noise pollution status, water contamination, energy consumption, etc. can also be estimated smartly in the smart city scenario. Therefore, environmental pollution monitoring, industrial pollution, household pollution monitoring, LPG leakage, food freshness etc. must also be smartly available in the smart cities. Therefore, the intelligent gas monitoring system is required in the smart city scenario.

1.2 Gas Monitoring

Emerging applications of Gas sensing attract both researchers and industrialists for its use in state-of-the-art areas. Humans are exposed to variety of gases/odors in their daily lives. Some of these gases/odors may have poisonous nature, both in short and long terms. Therefore, efficient gas monitoring is essential for safety of mankind and society. Several gas sensors are hence fabricated using variety of technologies and phenomena for gas sensing [4]. It has been found and reported by past researchers that an individual sensor may responds to multiple chemicals and are non-selective by nature, although the sensitivities of each sensor element may be different for different gases/odors [5].

Further, ambience consists of mixtures of gases/odors instead of single gas/odor although in such mixtures a few components may supersede making it single gas/binary/ternary mixture etc. Previous researchers observed that human olfactory system consists of about a thousand olfactory receptor genes forming millions of grouped to sense various smells. Thereafter, mimicking of human olfaction took place. Therefore, sensor array was being used instead of single sensor [6]. Introduction of pattern recognition techniques and artificial intelligence completely mimic the human nose into a machine olfaction and the terms "electronics nose (e-nose)" came into picture [6, 7]. It is a system having four basic components viz., sensor array, data processing, pattern recognition, and classification. There were several bottlenecks within these elements of the e-nose and further improvements were required. Sensor arrays generate patterns which are very complex in nature, so data processing is required to unveil the subtle information [8]. Therefore, a suitable data processing not only elevates the overall performance, it also reduces the complexity of entire e-nose system [9].

2 Intelligent Gas Monitoring

Gas sensing is being researched from almost three hundred years. The increasing applications in variety of the areas and the advancement in the science and technologies leads this as the prevailing area of recent research. There are two major aspects of gas monitoring. One is 'classifying' the target gas and other is 'quantifying' the amount of the target gas. Intelligent gas monitoring is the technology through which gas monitoring is achieved in real-time. The Machine Learning (ML) and Artificial Neural Network (ANN) technologies are helpful to implement intelligent gas monitors [10].

2.1   Gas Sensors

There were several gas sensors developed by various previous researchers and mostly they are mechanical sensors. In 1952, Brattain et al. experimented on Germanium (Ge). They were curious to see the variations in the contact potential of Ge. They experimented different size of Ge in different ambient condition. They observed that the electrical properties of semiconductors changes according to the gaseous ambience. They have performed various experiments on Ge to see the variations in the contact potential according to gaseous ambient [11]. This experiment was the main breakthrough in the area of electronics gas sensors. After this novel finding, many of the researchers tried to explore the gas sensing technique and got patents as well. There are various types of gas sensors available now-a-days. Depending upon the sensing mechanism previous researchers discriminated the gas sensing techniques in many ways. Among them the popular techniques are Metal Oxide Semiconductor (MOS) based sensors, Polymer based sensors, Carbon Nanotubes based sensors, Inorganic materials based sensors etc. which are utilizing the electrical property variations [4]. There are several other methods also used in gas sensing technologies. Such as, Optical Methods, Acoustic Methods, Gas Chromatograph Calorimetric Methods etc. Many performance metrics are used to evaluate the gas sensors like, sensitivity, selectivity, response time, energy consumption, reversibility, adsorptive capacity, and fabrication cost etc. MOS sensors are considered as most common and cost effective gas sensors with considerably better performance in sensing applications [4, 12]. These sensors are based on electrical property variations. Figure 1 shows the MOS based gas sensor element with its electrical equivalent. The sensing mechanism in the MOS gas sensor involve redox reactions. Therefore, these gas sensors are required heating for sensing gases.

As shown in Fig. 1, the MOS sensor has sensing material, electrodes, substrate, and a heater. The sensing material is responsible to sense the gas and the response corresponding to the amount of interacting gas is observed from the variation in potential of the electrodes. The substrate is to support the gas sensor only, that is why

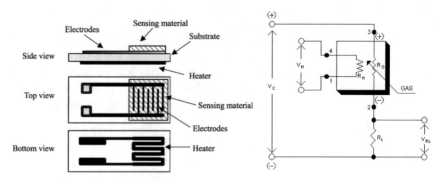

Fig. 1 Typical structure of a MOS gas sensor and its electrical equivalent circuit [12]

it is usually fabricated by using alumina or the glass. Heater patterns are fabricated by metals to provide sufficient temperature for maintaining the redox reactions. It has very simple working principle. The interacting gas molecules tends to vary the resistance of its sensing element by the phenomenon of physisorption. It is the short form of physical adsorption.

2.2 Bottlenecks in Gas Monitoring

Gas monitoring comprised of two important tasks, classification and quantification. Classification of gases/odors provide the detection of any particular gas present in the target sample. Quantification process provide the amount of that gas/odor in the sample. Monitoring of gas requires both these task perfectly [13]. The gas sensor is the key element for gas monitoring. Therefore, most of the bottlenecks are associated to the gas sensors itself. There are many performance indicators and bottlenecks associated with the gas sensor element [4, 6, 8]. Only the most popular performance indicators and corresponding bottlenecks associated to the gas sensors are listed below.

Most popular performance indicators

- Sensitivity: the minimum amount of target gas that can be detected
- Selectivity: the ability to detect only specific gas
- Response time: the time period to achieve steady state
- Recovery time: the time period to return the baseline.

Corresponding bottlenecks with the gas sensor

- It should be highly sensitivity (i.e., it can be able to sense very small amount of target gas even)
- It should be highly selectivity for specific gas
- Its response time should be higher
- Its recovery time should be lower.

We have seen that the redox reactions are responsible for the sensing of gases. Therefore, 'selectivity' is the most common issue associated with the gas sensors. Mostly the MOS sensors are facing the problem of non-selectivity because they respond for bunch of gases (instead of the specific gas) which can participate in the redox reaction. Use of array of sensing elements in place of the single sensor element is better solution to avoid the problem of non-selectivity. Therefore, the use of multiple sensor elements forms a gas sensor array. Each of individual sensor element of such array respond differently when they are exposed to any gas sample of particular concentration. Therefore, we have some pattern with the help different sensor element response. Hence a unique signature pattern of that gas sample is generated by using the array. Figure 2 show the four sensor element MOS gas sensor array along with the single sensor element.

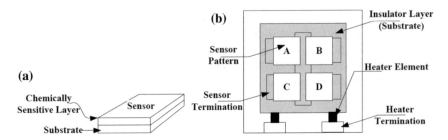

Fig. 2 a Individual gas sensor element. **b** Top-view of four element gas sensor array [14]

2.3   Intelligence Gas Monitoring

The machine learning and artificial neural network technology help to obtain the intelligent gas monitoring. Recently, researchers are trying to develop more accurate gas detection algorithms by applying the ANN and ML. Real-time gas monitoring is the main objective for the recent research in this area. The IoT technology with gas sensor arrays, signal processing, and machine learning techniques changed the classical gas monitoring to intelligent gas monitoring. Now, smart homes, smart wearable and handheld gazettes, etc. have been developed by utilizing the intelligent gas monitoring [10]. These technologies and devices are suitable to the smart city scenario. In 1982, Dodd and Persaud suggested the idea of intelligent gas sensing and coined the name 'Electronic Nose (Enose) [7]. Thereafter, many of the progress have been reported in this area. Basically, there are three main elements in the intelligent gas sensing system as shown below in Fig. 3 [15]. As the performance of intelligent gas sensor system depends upon these elements. Each of these elements has variety of options for researchers. Also, suitable data preprocessing along with the ANN algorithms can enhance the performance of overall system drastically. Therefore, data preprocessing step has attracted many of the researches.

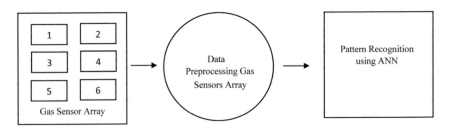

Fig. 3   Basic elements of intelligent gas sensor

2.3.1 Data Preprocessing in Intelligence Gas Monitoring

As discussed earlier, there are many limitations associated to the gas sensors such as; sensitivity, selectivity and stability [4, 16]. Even the utilization of sensor array also has poor performance in many cases because of the complex signatures of gases/odors. The performance of sensor array can be improved further by transforming their responses with some suitable data pre-processing [9, 14, 16, 17]. Further, the data preprocessing also associates the samples same gas having different concentrations to provide better cluster. Data processing algorithms have various purposes. It is used as signal processing for noise removal and signal conditioning. Again data preprocessing is used for normalization as well as to extract the desirable information. This information is popularly called as features. Generally, the raw sensor responses from the sensors are noisy, and signatures contained by these raw sensor responses are mostly subtle and quite complex in nature. ML and ANN algorithms cannot perform good classification results by directly analyzing these signatures. Therefore, data preprocessing is required. It is suggested by the researchers that pre-processing of the raw data can immensely augment high efficiency to the classification results. Actually, the data pre-processing transforms raw sensor responses to some newer analysis space. This transformation unveils the subtle information in the newer analysis space. Gardner et al., suggested several data preprocessing algorithms. Thereafter, many researchers suggested various remarkable data preprocessing algorithms. Recently, a robust data preprocessing technique has been reported for intelligent gas sensing. It is called as Normalized Difference Sensor Response Transformation (NDSRT) [18]. Details of the NDSRT will be discussed in next section. Beside of several advantages there are some drawbacks associate to the data preprocessing in context of gas monitoring. As we know the gas monitoring has two task, classification and quantification. Also, we found that data preprocessing helps the classification task and enhance the classification performance of intelligent gas monitoring system. In contrast to classification, the data preprocessing degrades the quantification performance. We will discuss this issue in detail in forthcoming section.

2.3.2 NDSRT

Mishra et al., published an efficient transformation called Normalized Difference Sensor Response Transformation (NDSRT) in 2017 [18]. According to them, it is a difference based normalization transformation which produce virtual sensor responses of the target gas samples. They suggested that variation in the concentration is responsible for the misclassification and poor classification performance in intelligent gas sensing. The NDSRT transformation produce almost concentration independent responses. Therefore, the virtual multisensory responses obtained by NDSRT show crisply identifiable clusters. According to them, principal component analysis (PCA) and other well established transformation or normalization techniques are unable to generate well separable clusters. They used the responses of four-element thick film gas sensor array for four obnoxious gases to show the efficacy of the NDSRT. Their result shows an average cluster compaction of 94.38% as compared to the PCA. Also, the non-overlapping clusters were obtained only by NDSRT. The definition of NDSRT is given below.

If there is a sensor array consist of 'n' gas sensor elements. Then the sensor array response vector for any gas sample will be 'X'. It has 'n' number of responses corresponding to each sensor element. It can be defined as [18]:

$$X = \{x_1, x_2, \ldots, x_i, \ldots, x_j, \ldots, x_n\}$$

Here, 'x_i', and 'x_j' are the response of 'ith' and 'jth' sensor elements respectively.

Therefore, the NDSRT response will be,

$$\text{NDSRT} = \frac{(x_i - x_j)}{(x_i + x_j)} \ \forall i < j$$

Therefore, the number of virtual sensor '$V.S.$' responses generated by NDSRT will be,

$$V.S. = n_{C_2} = \frac{n(n-1)}{2}$$

Figure 4 renders clustering through NDSRT transformation. It is obvious from the figure that the raw sensor responses are containing concentration information also. That is why if we see the scatter formed by the raw responses then it will be the overlapping and expanded clusters. In contrast, the NDSRT responses (virtual sensor responses) are not so much affected by the concentration of the gas sample.

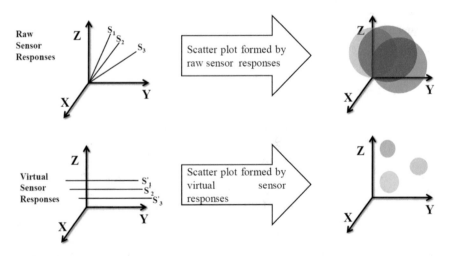

Fig. 4 Illustration of the NDSRT transformed clusters

The scatter plot formed by these virtual sensor responses has well separated and compact clusters. Therefore, NDSRT becomes robust data preprocessing technique and forms good clusters.

2.3.3 Classification in Intelligence Gas Monitoring

Good clusters support better classification. Cluster is basically grouping similar objects. Therefore, there will be as good and compact clusters as less variation in the characteristics of similar objects. In context of intelligent gas sensing, the varying concentration only introduce the variation in the characteristics of samples of same gas/odor. This variation can be suppressed by removing the concentration information only. Accordingly, the data preprocessing (transformation) can be utilized to enhances the similarity and suppresses differences among the gas samples. Also, the transformed responses support better classification. As it can provide closer association of different samples of the same class of gas. Also, it helps in reducing the complexity of the classifiers. ANN are considered as efficient classifiers in intelligent gas monitoring [19]. Multi-layer perceptron (MLP) is a considered as very classifier in case of intelligent gas monitoring [13, 14]. It consists of feedforward artificial neural network architecture. The MLP classifier has basically three layers commonly known as input layer, hidden layer, and output layer. Figure 5. shows the simpler MLP classifier architecture for intelligent gas monitoring. It utilizes the backpropagation algorithm to evaluate the errors for efficient training [20]. The ANN classifier performance depends greatly upon the fed input. Therefore, ANN complexity can be reduced if well preprocessed data is fed as the input. Further, we can have real-time gas classification by utilizing the ANN for the data preprocessing also.

Real-time ANN classifier for intelligent gas monitoring is shown in Fig. 6. It has two ANN stages consists of the MLP architecture for data preprocessing and other has simpler MLP classifier for intelligent gas monitoring.

Fig. 5 A simpler MLP classifier for intelligent gas monitoring

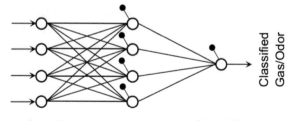

Input Layer   Hidden Layer   Output Layer

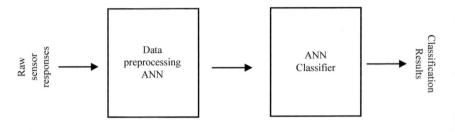

Classifier in intelligent gas monitoring

Fig. 6 Real-time classifier in intelligent gas monitoring

2.3.4 Quantification in Intelligence Gas Monitoring

Quantification is another important task of the intelligent gas monitoring. In contrast to the classification task, it actually requires the corresponding variation in the concentration of corresponding gas samples. The gas sensor responses should vary with any change in concentration to achieve accurate quantification. We have already seen in previous sections that the concentration effect is altered in the pre-processed data. Therefore, certainly this preprocessed data will provide inaccurate quantification results. Therefore, it is recommended that the data preprocessing only supports the classification but, adversely affect the quantification results [13]. The effect of preprocessed data can be observed in many previous researches. In 2010, Ravi et al. [21] and in 2013, Sharma et al. [22] have shown that data preprocessing supported better classification results. However, the quantification results have been found very poor. Figure 7 depicts a simpler quantifier ANN architecture using feed-forward back-propagation algorithm for intelligent gas monitoring. Further enhancement in the performance of the quantifier can be achieved by utilizing the parallel ANN architectures [13]. Mishra et al., in 2018, reported that the performance of quantifier can be improved by using the raw sensor responses directly [13]. Further improvement can be achieved by using parallel architecture [13]. In parallel architecture different ANNs are utilized to quantify corresponding class only. Therefore, the improvement is achieved at the cost of complexity. In this scheme the same number of different ANNs are required as there are target classes of the gas/odor.

Fig. 7 Simpler quantifier ANN architecture in intelligent gas monitoring

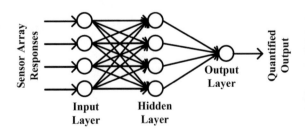

Fig. 8 Parallel quantifier
architecture in intelligent gas
monitoring

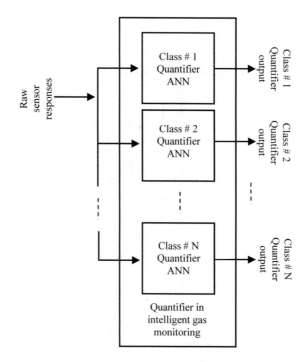

The illustration of such parallel quantifier architecture is shown in Fig. 8. Here, it is observed that there are '*N*' number of parallel quantifier ANNs corresponding to the '*N*' classes of gases/odors. Each of these quantifier ANNs has different ANN architectures. These ANNs have very simpler architecture similar to the architecture given in Fig. 7. We will understand the benefit of parallelism in the quantifier ANN architecture in next section.

2.3.5 Modular ANN Architecture in Intelligence Gas Monitoring

It consists of both classifier ANN and quantifier ANN for intelligent gas monitoring. The modular is given because it can be scaled according to the requirement. Figure 9 renders an example for modular ANN architecture for intelligent gas monitoring [13]. In this modular ANN architecture there are two ANN blocks namely, classifier ANN block and quantifier ANN block. Further, the classifier ANN block has again two neural network architectures. One is destined for data preprocessing and named as $Net_{1\text{-}NDSRT}$ and another one is designed for classification purpose. It is named as $Net_{2\text{-}classifier}$. The raw sensor responses are fed to the first neural network for data preprocessing.

Then the processed data is available as the input for next neural network which actually is designed to classify the gases/odors. In this modular architecture there are four classes (gases). Therefore, two neurons are sufficient to provide the results

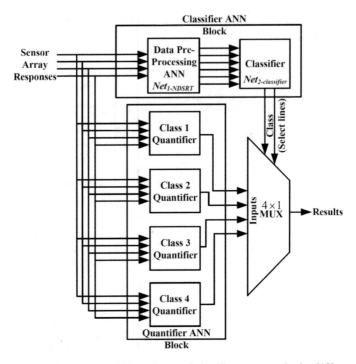

Fig. 9 An example for modular ANN architecture in intelligent gas monitoring [13]

in binary format. Now the same raw sensor responses are available at the quantifier ANN block also. It is fed parallel to each of the parallel quantifiers. Therefore, each of these parallel quantifier ANN architecture have the corresponding quantified results, simultaneously. Therefore, the correctly classified gas has its corresponding concentration through the multiplexer which is used to multiplex the results of classifier ANN block and the quantifier ANN block. Therefore, the parallel ANN architecture has several merits. This architecture has scalability, faster output, and this architecture provide the further improvement in the quantification results. Therefore, the concept of utilizing modular ANN architecture in intelligent gas monitoring supports the smart city scenario.

3 Blockchain for Intelligent Gas Monitoring

In recent years, blockchain has gained lot of attention from different fields such as academia, industry, finitech, companies, automotive, IoT industry and so on [23, 24]. Blockchain has broad range of potential that creates remarkable impact on IoT industry. Blockchain is an effective technology for creating trust and security in IoT fields, where the IoT sensor nodes do not trust each other. The IoT systems need to trust the data gathered from various sensors to ensure end-to-end data integrity and security. The gathered data from the sensors should be pre-processed, analyzed and securely stored in distributed ledger such as blockchain so that the malicious attackers cannot tamper the data. In addition to this, blockchain provides auditability and transparency of the stored data so that any member of the blockchain network can verify the stored sensor data giving more trust to the stored data. Moreover, the blockchain can be integrated with different types of sensor devices in the smart city that can provide a shared platform. All the IoT devices connected with the blockchain network will be able to communicate and exchange information with each other in a secure way. The collected information is logged in the database for unlimited time as well as stored data can be used for auditing.

3.1 Blockchain

Blockchain is an emerging distributed and decentralized technology that underpins the most dominated cryptocurrency called Bitcoin. An unknown person or a group known by the name Satoshi Nakamoto first introduced the popular cryptocurrency Bitcoin [25]. The popularity behind blockchain is due to its long list of advantages. Some of the advantages of blockchain are as follows. The blockchain is a distributed and decentralized technology that prevents from single point of failure attack, which is the major problem in centralized system. The data stored in the blockchain is more accurate, consistent and has great transparency, as it is available to all the members of the blockchain networks who have permission to access. It is very difficult to change a single data stored in the blockchain, as it requires huge amount of work and energy. Since, the data are stored in a distributed manner, i.e. all the member nodes stored the data instead of storing in a centralized server, it is very hard for the attacker to compromise the data stored in the blockchain [26]. In case of traceability, it is easy to trace the data in blockchain as the historical data are stored in the blockchain in chronological order and it is easy to verify the authenticity of the stored data and can prevent fraud. Since, there is no intermediary in case of blockchain, it reduces the cost that is being spent for the brokers, for example advertising agency. It also provides partial anonymity and privacy to the users as only the stored data or transaction are public but the identity are not.

A blockchain is a distributed shared ledger that stores chain of blocks with data in chronological order. The chaining of blocks is performed by addition of the previous block's hash to the current block, then the current block's hash to the next block in a sequential manner as shown in Fig. 10. Then, it is shared with other nodes in a distributed peer-to-peer networks in a secure way without the need for a central authority. The consecutive hashes of blocks guarantee transaction to come in a chronological order. This prevents the blockchain from double spending attack or 51% attack [27]. The blockchain is protected by consensus mechanism, i.e. all the nodes of the blockchain network agree on a common judgement or answer. In case of Bitcoin, the consensus mechanism is Proof of Work (PoW). A Proof of work (PoW)

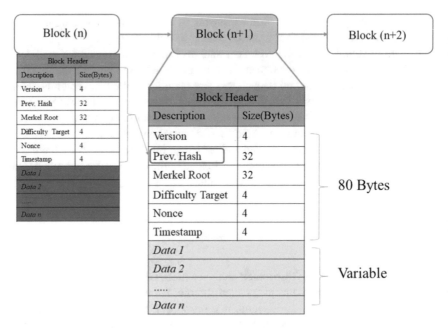

Fig. 10 Blockchain working principle

is a mathematical puzzle that is very hard to solve hence securing the blockchain double spending attack. It is very hard to find a nonce in proof of work but once found it is easy to validate the new block. The blockchain is verified by consensus of anonymous nodes in generation of blocks. The blockchain is considered secure, as the collective computational power of the malicious nodes does not dominate the computational power of the good nodes [28]. The PoW consensus mechanism verifies that the miner is not influencing the network to make bogus blocks.

3.2   Intelligent Gas Monitoring Based on Blockchain

The intelligent gas monitor is one of the IoT sensor devices that is part of smart city. The intelligent gas monitoring consists of smart sensors and communicating devices to monitor the poisonous gas, gas leakage etc. in the smart city. The gas sensors are wirelessly connected to the IoT network. The sensors collect the sensed data then transmit, receive or stored the data first in the local sensor node and then to the IoT network. Security requires highest priority in case of distributed sensors as it is very easy to tamper isolated devices or the collected data sensed by the sensors using man-in-the-middle attack. If the data is tampered or fake information is circulated, then it might create a havoc in the smart city. Hence, it is important to provide security to the stored data and it can be done by using blockchain technology.

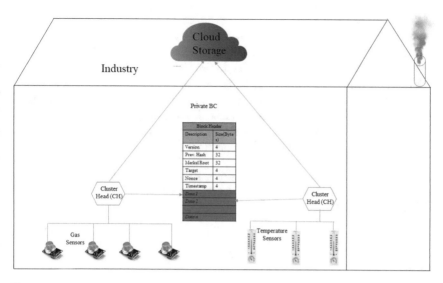

Fig. 11 Blockchain assisted intelligent gas monitoring

We consider a typical chemical industry (or gas industry) where various experiments are done by combining different types of chemicals and some of the chemical products might be poisonous. The industry is equipped with several IoT devices including intelligent gas monitors with sensors as shown in Fig. 11. In this section, we will discuss on how the sensor data generated by the intelligent gas monitoring devices are stored in blockchain. Unlike Bitcoin where the blockchain is a public blockchain and all the transaction are stored publicly, we consider a private blockchain as it is secure and suitable for IoT environment. All the IoT devices should get permission to be the part of private blockchain. In this case, we consider the collected sensor data as transactions. The blockchain is managed and controlled by a private owner, where he can add or delete the IoT devices by deleting from its ledger. The private blockchain will have a policy header that acts as an access control list. The owner of the blockchain has authority to control all the data transaction happening in the industry. Since, a private blockchain is used, there is no need to use expensive consensus mechanism like PoW, so the excess overhead can be minimized easily. A simple Practical Byzantine Fault Tolerant (PBFT) consensus mechanism can be used. A Cluster Head (CH) will collect all the sensor data collected by each type of sensor devices like gas sensors, temperature sensors etc. For example, the intelligent gas monitors have their own cluster head, likewise the temperature sensor have their own CH. The collected data from the sensors can be stored locally in CH, shared in the IoT network, or can be stored in the cloud. The CH acts as a verifier and after analyzing the data, the CH can propose a new block in the network. The CH accept or discard the sensor data based on the authenticity and integrity of data. After collecting and analyzing the sensor data, the CH generates a new block for the blockchain along with multi-signature transaction to its neighboring CH. The CH sends the newly created block and the multi-signature to the neighbor CH and the neighbor CH tries to verify the received multi-signature transaction. Here, if all the CH, i.e. verifier validates and accepts the new block, then only the block is added in the blockchain as the process mentioned in Sect. 3.1.

Hence, intelligent gas monitoring based on blockchain handles privacy as well as security attacks such as DoS attack, message modification attacks, mining attacks etc., while considering resource constrained IoT sensors. The malicious attackers cannot tamper the sensor data stored in the blockchain easily. The distributed storage of the data helps to perform intense computation efficiently.

Acknowledgements This research was supported by Korea Research Fellowship program funded by the Ministry of Science and ICT through the National Research Foundation of Korea(NRF-2019H1D3A1A01071115).

References

1. Harrison C, Donnelly IA (2011) A theory of smart cities. In: Proceedings of the 55th annual meeting of the ISSS-2011, Hull, UK, vol 55, No 1
2. Ismagilova E, Hughes L, Dwivedi YK, Raman KR (2019) Smart cities: advances in research—an information systems perspective. Int J Inf Manage 47:88–100
3. Deakin M, Waer H (2011) From intelligent to smart cities. Intell Build Int 3(3):140–152
4. Liu X, Cheng S, Liu H, Hu S, Zhang D, Ning H (2012) A survey on gas sensing technology. Sensors 12(7):9635–9665
5. Ponzoni A, Baratto C, Cattabiani N, Falasconi M, Galstyan V, Nunez-Carmona E, Rigoni F, Sberveglieri V, Zambotti G, Zappa D (2017) Metal oxide gas sensors, a survey of selectivity issues addressed at the SENSOR Lab, Brescia (Italy). Sensors 17(4):714
6. Gardner JW, Bartlett PN (1994) A brief history of electronic noses. Sens Actuat B: Chem 18(1–3):210–211
7. Persaud K, Dodd G (1982) Analysis of discrimination mechanisms in the mammalian olfactory system using a model nose. Nature 299(5881):352
8. Zarcomb S (1984) Theoretical basis for identification and measurement of air contaminants using an array of sensors having partially overlapping sensitivities. Sens Actuators 6:225–243
9. Gutierrez-Osuna R, Nagle HT (1999) A method for evaluating data-preprocessing techniques for odour classification with an array of gas sensors. IEEE Trans Syst Man Cybern Part B (Cybern) 29(5):626–632
10. Feng S, Farha F, Li Q, Wan Y, Xu Y, Zhang T, Ning H (2019) Review on smart gas sensing technology. Sensors 19(17):3760
11. Brattain WH, Bardeen J (1953) Surface properties of germanium. Bell Syst Tech J 32(1):1–41
12. Arshak K, Moore E, Lyons GM, Harris J, Clifford S (2004) A review of gas sensors employed in electronic nose applications. Sensor Rev 24(2):181–198
13. Mishra A, Rajput NS (2018) A novel modular ANN architecture for efficient monitoring of gases/odours in real-time. Mater Res Exp 5(4):045904
14. Mishra A, Rajput NS, Singh D (2018) Performance evaluation of normalized difference based classifier for efficient discrimination of volatile organic compounds. Mater Res Exp 5(9):095901
15. Di Natale C, Davide F, D'Amico A (1995) Pattern recognition in gas sensing: well-stated techniques and advances. Sens Actuat B: Chem 23(2–3):111–118
16. Korotcenkov G, Cho BK (2017) Metal oxide composites in conductometric gas sensors: achievements and challenges. Sens Actuat B: Chem 244:182–210

17. Rajput NS, Das RR, Mishra VN, Singh KP, Dwivedi R (2010) A neural net implementation of SPCA pre-processor for gas/odor classification using the responses of thick film gas sensor array. Sens Actuat B: Chem 148(2):550–558

18. Mishra A, Rajput NS, Han G (2017) NDSRT: an efficient virtual multi-sensor response transformation for classification of gases/odors. IEEE Sens J 17(11):3416–3421

19. Zhang GP (2000) Neural networks for classification: a survey. IEEE Trans Syst Man Cybern Part C (Appl Rev) 30(4):451–462

20. Kermani BG, Schiffman SS, Nagle HT (2005) Performance of the Levenberg–Marquardt neural network training method in electronic nose applications. Sens Actuat B: Chem 110(1):13–22

21. Kumar R, Das RR, Mishra VN, Dwivedi R (2010) A neuro-fuzzy classifier-cum-quantifier for analysis of alcohols and alcoholic beverages using responses of thick-film tin oxide gas sensor array. IEEE Sens J 10(9):1461–1468

22. Sharma S, Mishra VN, Dwivedi R, Das RR (2013) Quantification of individual gases/odors using dynamic responses of gas sensor array with ASM feature technique. IEEE Sens J 14(4):1006–1011

23. Kim S, Deka GC (2019) Advanced applications of blockchain technology

24. Kim S, Deka GC, Peng Z (2019) Role of blockchain technology in IoT applications. In: Advances in computers

25. Nakamoto S (2016) Bitcoin: a peer-to-peer electronic cash system, December 2008

26. Shrestha R, Bajracharya R, Shrestha AP, Nam SY (2019) A new-type of blockchain for secure message exchange in VANET. Digit Commun Netw

27. Shrestha R, Nam SY (2019) Regional blockchain for vehicular networks to prevent 51% attacks. IEEE Access 7:95021–95033

28. Aitzhan NZ, Svetinovic D (2016) Security and privacy in decentralized energy trading through multi-signatures, blockchain and anonymous messaging streams. IEEE Trans Dependable Secure Comput 15(5):840–852

Commodity Ecology: From Smart Cities to Smart Regions Via a Blockchain-Based Virtual Community Platform for Ecological Design in Choosing All Materials and Wastes

Mark D. Whitaker and Pravin Pawar

Abstract This article describes three points: (1) how a fine-grained yet systematic model of regional material sustainability called Commodity Ecology is being linked to current technological trends (2) of societies evenly saturated with ever cheaper and more mobile Information Communication Technologies (ICT) and (3) increasingly organized by decentralized ledgers like Blockchain. We live in a society connected by the Internet and increasingly served by and even ruled by online platforms. Particularly, an evenly shared mobile-phone revolution in ICT is putting platform access in the hands of all peoples around the world. Another disruptive technology after the Internet is blockchain technology which has revolutionized the exchange of information in areas such as cryptocurrency, supply chain management, healthcare and smart contracts. Thanks to these technological developments, instead of just smart cities, we now live among real/virtual 'smart regions' that on their own are capable of deliberating, finding, and buying/selling en masse toward their own better choices in sustainable material choices and better waste handling. These trends are utilized by the first online blockchain-based virtual community platform for 'smart regions' that facilitates this global, multi-regional drive to more democratic, holistic, and sustainable ecological design in all of our consumptive choices. Commodity Ecology is a model, rubric, and checklist for sustainability in 130 material categories, on a virtual platform for deliberation in over 860 distinct ecoregions simultaneously. The United Nations Academic Impact Office called Commodity Ecology a top global initiative for actually achieving Sustainable Development Goal #12, Responsible Consumption and Production.

M. D. Whitaker (✉)
Department of Technology and Society, Stony Brook University, State University of New York, Korea, Incheon, Songdo, South Korea
e-mail: Mark.Whitaker@sunykorea.ac.kr

P. Pawar
Department of Computer Science, Stony Brook University, State University of New York, Korea, Incheon, Songdo, South Korea
e-mail: Pravin.Pawar@sunykorea.ac.kr

M. D. Whitaker · P. Pawar
SUNY Korea, Incheon, South Korea

© Springer Nature Singapore Pte Ltd. 2020
D. Singh and N. S. Rajput (eds.), *Blockchain Technology for Smart Cities*,
Blockchain Technologies, https://doi.org/10.1007/978-981-15-2205-5_4

Keywords Commodity ecology · Blockchain · Smart region

> If we are to create a sustainable world...we must recognize that our present forms of agriculture, architecture, engineering, and technology are deeply flawed. To create a sustainable world, we must transform these practices. We must infuse the design of products, buildings, and landscapes with a rich and detailed understanding of ecology....They can be solved only if industrial designers talk to biogeochemists, sanitation engineers to wetland biologists, architects to physicists, and farmers to ecologists. In order to successfully integrate ecology and design, we must mirror nature's deep interconnections in our own epistemology of design.
>
> —Sim Van der Ryn and Stuart Cowan,
>
> in *Ecological Design, Tenth Anniversary Edition*

1 Introduction

1.1 Our New "Network Society" Creates the Possibility for "Smart Regions" Via Societies Saturated with Mobile ICT, Equalized by Common Online Platform Participation and Powered by Blockchain Revolution

A different deliberative and regionalized future is being created in the early 21st century in our society of equitable mass communication. Changes of media regimes have been neglected in understanding changing political economic dynamics in world history [1] Today's change of media regime toward living in societies characterized by saturated mobile ICT networks is making deeply deliberative 'smart regions' possible for the first time. This has never happened before. One has to go back to the beginning of speech to see any communications regime with such an equalizing and localizing potential. Plus, our current digital electronic technologies have little problems of long-term storage or long-distance communication. This gives almost anyone capacity to organize large and equitably participative projects of 'big data' because of such widely 'digitally enfranchised' peoples. The online platform for Commodity Ecology is such a project.

This new media society has been called the "network society" [2] because even though we have always lived in networks, we are now living in microelectronic-enhanced networks that alter the way we organize networks in four ways:

- **Flexibility**: electronic networks can reconfigure "according to changing environments, keeping their goals while changing their components. They go around blocking points in communication channels to find new connections." Past networks lacked such capacities [3].
- **Scalability**: electronic networks "can expand or shrink in size with little disruption." It was very costly once to expand networks in the past, though once a

particular platform is set up now, scalability of electronic networks can be global rather easily [3]. In a blockchain based network, any node can join and leave at will facilitating the network scalability even further.

- **Transparency:** Thanks to the blockchain technology revolution, electronic networks can be fully transparent and open their databases to all involved. All the nodes connected through the internet maintain all of the transactions made on a blockchain network collaboratively. The shared ledger is updated every time a transaction is made. It is publicly available and is incorruptible which introduces transparency to the system [4].
- **Survivability**: "[B]ecause they have no center," (blockchain-enabled) electronic networks "can operate in a wide range of configurations, and networks can resist attacks on their nodes and codes because the codes of the network are contained in multiple nodes that can reproduce the instructions and find new ways to perform." The database in a blockchain-based network is composed of blocks of information which is copied to every node of the system. Each block contains a list of transactions and a timestamp which links it to the previous chain of the blocks [4]. "So, only the physical ability to destroy the connecting points can eliminate the network." Older networks were easily disrupted in such ongoing deliberations, while current electronic networks find such disruptions just a challenge of reorganization instead of an existential crisis [3].

Because of the above points, think how different are political economic and cultural possibilities now. These possibilities can be contrasted with a past regime of communications that is now being challenged by the current "network society." To conceptualize this change, think of this change of media regime into our current saturated mobile ICT networks as a move away from a 1-way mass media broadcasting (with expensive technologies only allowing for powerful central voices and masses of silent viewers/listeners) into 2-way mass media broadcasting (with ever cheaper distributed technologies allowing greater equality of capacities to speak, to create/produce media, to respond, and thus to deliberate between all nodes). This has been created by ever cheaper, mobile, smaller, networkable, distributed communications technology. Because of these trends, it is estimated by January 2019 there are over 5.1 billion mobile phones, with more than half (55%) of those mobile phones already smartphones—which allows for the majority (52%) of total Internet access globally to be done via smartphones now [5]. Because of this, three additional points about this "network society" can be demonstrated that are leading us culturally to live in self-deliberating 'smart regions':

- Easier *self-organized networks of flexible communities* fail to require elite sponsors.
- There are *radically reduced urban-to-rural inequalities in communication* for the first time in world history, which influences development and political participation. This means:
- *Novel parities in the 'means of communication'* are occurring now for only the second time in world history. The first time was the invention of language.

We can conceive of ourselves living in a major time of "media/society mismatch", in which old institutions built within principles of a previous media regime fail to fit the faster, deliberative, changing political dynamics of our current media regime. This has been poetically summarized by Argentina's software designer Pia Mancini, another designer of online platforms for mass deliberation (as inventor of "Democracy OS"): "We are 21st-century citizens, doing our very best to interact with 19th century-designed institutions that are based on an information technology of the 15th century" [6]. Similarly, it is argued that an inherited debate from the 1990s about the difficult-to-define term "smart cities" [7]—which generally means how to create more sustainable, deliberative cities with ICT—equally will be outmoded by a wider debate about "smart regions": how to organize a sustainable project far wider, more participative, and more ecologically holistic in conception than just an urban node.

In short, mobile ICT networks are rewiring social life rapidly and changing the background social context of our past formal institutions linked to different media regimes. This network society is rewiring state politics, journalism, knowledge access, consumption, and financial currencies. This is because the current media regime has the potentialities of creating very powerful self-organizing grassroots/community development via mobile networks 'from below.' These grassroots trends appealed to and unified by Commodity Ecology are highlighted by italics below:

- State: 'Netizens' exist now that are *heard from easily everyday* far more than citizens polled intermittently at election times. *Online political campaigns can reach millions and even billions easily, and they can organize across regions quickly on neglected issues. There are novel kinds of free and participative journalism.* From above, there is state erosion of privacy. *From below, there is a desire and a concern for greater encrypted privacy.*
- Science/Knowledge Transmission: *The social media rate globally in 2019 is now at 3.5 billion people* [8], *allowing 2-way mass communication of news and information without filters of 1-way media companies or states setting narratives, goals, or priorities of our societies alone.* There is additionally much *online education,* increasing rule by biased algorithms [9], *searchable world knowledge for everyone* for the first time, global spying based on 'big data,' and 'shadowbanned' individuals and 'de-platformed' dissidents who want to make dissident platforms in turn. Our 2-way mass media is a multi-directional, unpredictable and viral route of news and information. *Decentralized databases like Blockchain (from 2009-2010) and Holochain (only in the past few years) were invented to find solutions to get beyond past 'primary node' third-party monitoring and recording of transactions, in order to find some way to employ the network logic of culture and design where all network nodes are equal and participative. Thus the desires for greater privacy yet greater deliberation and transparency are being acted upon equally.*
- Consumption: This is the *material, laboring, productive, consumptive, and pollutive aspect of societies.* The network society is seeing the creation of the "sharing economy," *large platforms for online shopping,* 'prosumers' (who produce and consume equally and intermittently—instead of only having past centralized organizations produce and mass aggregates consume), the expansion of the 'gig economy' (into which some jump voluntarily and some are pushed because the latter

have become 'precariats' [10] without future options of retraining as past models of stable job positions of lifelong employment are decimated by competition from online choices in shopping as well as labor.) *Now we live in a world of large global e-shopping platforms and well as greater grassroots regional consumption options which attract easier awareness of their existence.*

- Finance: Our network society is seeing *the invention of "P2P" (peer to peer) cyptocurrencies, wider uses of regional community/complimentary currencies,* and online crowdfunding that gets around past gatekeeping of major financial banks on industrial level production. Especially, blockchain technology is seen as a path-breaking innovation and forerunner of a fresh economic period giving rise to a new type of system called the Blockchain Economic System [11]. In blockchain technology, the authenticity of a financial transaction is checked by the consensus protocol, which eliminates the involvement of a trusted third party for validation purpose [4].

2 Defining "Commodity Ecology"

2.1 A Merger of Holistic Ecological Design, ICT and Smart Regions

Commodity Ecology is an online platform that embraces our current media regime very well. The platform is:

1. A United Nations-recognized model [12] of sustainability comprising a taxonomy of 130 different categories of materials/technologies to facilitate innovation toward economically-sound and ecologically-sound development;
2. Operating in each of our world's 867 ecoregional zones; and
3. Available via an online software platform to implement the SDGs via network value of over 5 billion mobile phones as of 2019.

A list of these 130 commodity use categories organized symbolically as a wheel or as a straight list is available [13]. The "CEMVC" platform (the abbreviation of 'Commodity Ecology Mobile Virtual Community, already online as a prototype at comwheel.azurewebsites.net) merges greater local business incubation, greater local democratization, and greater sustainable development. It will improve society's future and create new sustainable businesses by organizing discussion via the abovementioned commodity taxonomy of 130 categories in each of 867 ecoregions of the world, crossing geographical and political boundaries to facilitate regional collaboration between consumers and producers. This project creates new value by combining ICT, humanities, social sciences, education, environment, culture and art by catalyzing an open public debate worldwide on how to design any "smart region's" greener, sustainable future.

2.2 Three Problems with Implementation of the U.N.'s Sustainable Development Goals (SDGs) that Commodity Ecology Solves

The concern with the development of the online version of the Commodity Ecology Platform is that it is presumed that the Sustainable Development Goals (SDGs) are in danger of falling short of being achieved by lacking a way to integrate 'people power' and local business incubation into their implementation in the long term, cheaply. As of this writing, even the United Nations is about to declare bankruptcy in late 2019, so relying on centralized institutions and centralized grants to achieve the "SDGs" seems a very old-fashioned way to think about implementing sustainability.

Seventeen global goals were agreed to by all nations at the UN General Assembly in 2015 to be fulfilled as a target by the year 2030. Before social media and a mobile phone-based culture, there was no cheap way to employ networks of real/virtual communities and online deliberation toward the SDGs and no way to encourage or catalyze all the regional differences of the world in their own enlightened self-interest toward these SDG goals. No virtual community existed to bring people together globally and regionally together in real communities on actionable projects toward this greater material sustainability for the short term much less a durable cultural long-term. Commodity Ecology is that novel way to organize sustainability for the long term, regionally across the world that takes advantage of over 5 billion mobile phones for cheap ongoing real/virtual deliberation. This is the Commodity Ecology model as an online platform.

This is why Commodity Ecology has been recognized by the United Nations Academic Impact Office (UNAI) as head of the list for implementing SDG #12 (Encourage responsible consumption and production patterns). Commodity Ecology provides three actionable levels that the current United Nations' Sustainable Development Goals ("SDGs") lack:

1. Commodity Ecology can implement the SDGs across a real world of disaggregated geography instead of abstract nations with completely different interpretations of success in these sustainable goals. Commodity Ecology is thus a clear and common rubric of geography. In this case, it is ecoregions that are currently chosen, already integrated into the platform as the ecologically-sound boundaries of deliberation. Ideally, other conceptions of tangibly real ecological regions can be integrated into the platform later, like watersheds. On the one

hand, the UN's SDGs assume abstract state territories as the rubric for sustainability and governments as the leading agency for that implementation. Both of these rubrics are unable to address holistically and ecologically boundary areas between states that have a great deal of common exclusion from the states in question, whether socially, economically or environmentally. This is because many countries choose rivers or water bodies as their political borders. However, such choices of easy transportation by water at a borderline only draw human development to border areas and leave poorly regulated borderland pollution and waste streams. However, on the other hand, ecoregions or watersheds as geographic metrics of sustainability avoid such issues because these rubrics place common water bodies at the center of the geographic zone for ecological design, instead of split across multiple jurisdictions of sustainability design. This rubric encourages cross-border collaboration as well, in a way that the SDGs are unable to do.

2. The second actionable level that Commodity Ecology has that the SDGs lack is clarity on the actual 130+ material categories of material sustainability. This clear and common rubric of 130 kinds of disaggregated materials is already integrated into the platform, and there can be nothing interpretive about that: people either have sustainable choices in all categories and integrate them together per region well, or they lack ideas for it. In other words, there is nothing interpretive about Commodity Ecology for how to achieve sustainability, while there is nothing except flexible interpretations and a lack of common standards when we look at how empty the SDGs are on how to avoid one country failing at sustainability yet coming to different definitions of success in an SDG category compared to another country. Commodity Ecology thus allows different regions to learn from each other using the same rubric since it is a common rubric—something that is less clear with the SDGs with different countries having different interpretations of these goals.

3. The third actionable level that Commodity Ecology has that the SDGs lack is how to reach disaggregated people in the billions cheaply in the long term. With the United Nations openly facing bankruptcy in late 2019 according to recent news, one can ask why we expect centralized administration or funding aid of ecological design projects in a network society. It is necessary or 'smart' development? To the contrary, in Commodity Ecology, reaching people cheaply and for the long term is already integrated into the mobile-accessible online platform since there are over 5 billion mobile phones rather evenly distributed in the world as of 2019. For instance, there is over 1 billion mobile phones in Africa by 2019 (with a population at 1.3 billion for a penetration at 80%), and another 1.2 billion mobile phones in India by 2019 (with a population 1.4 billion for a penetration at 87%). Therefore, mobile phone access is not suffering from a digital divide like other communication technologies [5, 8].

3 Development of Smart Cities into Smart Regions

The development of the concept of 'smart cities' and its interaction with sustainability and state development programs is well documented [14, 15]. However, what is less commented upon is the media context when this developmental idea of the 'smart city' was created. The concept of 'smart city' was conceived in the late 1990s when ICT still was relatively expensive and when the peer-to-peer potentials of network value of such services among the general public still was low. Thus higher transaction costs and central node solutions to sustainability kept making sense, and centralized solutions of 'smart cities' were used as a planning rubric for sustainability because that was the only expected major area served by ICT networks. With the rise of a smartphone saturated society, and wide rural coverage as well, this central node presumption at the heart of the 'smart city' has now ceased to be the case. Such a tiny urban-node-based definition of 'smart city' in all three senses of smart (smart ICT, smart infrastructure design, and ecologically smart) is untenable now. We have the luxury now to think on the level of 'smart regions' to be more holistic in our ecological designs instead of limited to urban-based ICT arrangements. As said earlier, this is because there is a growing parity across wide landscapes of rural and urban areas alike with over 20 years of growing mobile ICT coverage.

From 2015, the coverages of mobile phone connections have grown immensely [16]. For instance in that time all of Africa has moved from less than 100 million mobile phones to over 1 billion. In 2015, it was expected all of Africa would have around 500 million mobile phones by 2020. However, by 2019, Africa already has over 1 billion mobile phones—twice that amount which no one expected. The mostly rural extension of such mobile diffusion has been immense, and has called forth envisioning of 'smart regions' as now possible instead of only 'smart cities'. See Fig. 1 that depicts increasing saturation of electronic networks of smart regions as demonstrated by increasingly higher number of mobile connections compared to total population.

Thus, from these points, it is argued that past more centralized solutions for 'smart cities' now innately move toward "smart regions" as the developmental model because that is how we culturally live. Plus, it is argued that the model of 'smart regions' avoids the developed world bias that the concept of 'smart city' entailed as if 'smart development' can only occur in already industrially developed urbanized countries. Instead, it has been the case that diffusion of mobile phones has even leapfrogged electricity coverage in sub-Saharan African countries, and such a mobile service as a developmental platform is even used to develop more sustainable choices of how to develop and to fund electricity coverage under conditions of weak states and larger impoverished peoples [17]. These weak states that the UN's SDGs depend upon as a rubric will find it very hard to build roads much less implement SGDs and education toward it from a centralized administration. However, such "smart regions" worldwide are now developing. 'Smart regions' are the growing mobile network value of what previously was conceptualized as "ICT4SD" (Information Communication Technologies for Sustainable Development). ICT4SD

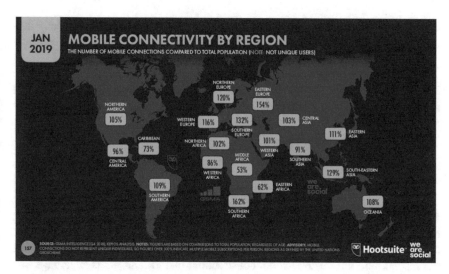

Fig. 1 Increasing Saturation of Electronic Networks of "Smart Regions" (Data from We Are Social, 2019)

is now as well influenced by the lowered transaction costs of regions of saturated mobile ICT networks, greater urban and rural parity of mobile phone coverages, and wider concern of environmental impact beyond mere urban sustainability. Equally, rural sustainability, reformation of industrial agriculture to other models, and much else about holistic material choices are in the offing. Think about plastic pollution. How can simply an urban based solution exist for such a problem? Instead, we require debate and deliberation from all involved about the act of choosing particular kinds of materials and their waste streams in general for the future. Commodity Ecology can provide such a platform and rubric of analysis that assures all categories of commodity use are being discussed in terms of their ecological design.

4 Details About the Commodity Ecology Mobile Virtual Platform (CEMVC)

A prototype of CEMVC already has been developed for this ongoing global collaboration aimed at an actual implementation of the SDGs capable of permanently reaching cheaply over 5 billion people particularly for less developed countries, instead of reaching them only sporadically with centralized grant-based programs. At this time of writing, the authors are seeking collaborators and research funding for completing the prototype for a truly globally robust rollout. Our objective is to improve a rough prototype of CEMVC under development and to take it to the global level with a secure multi-tiered architecture that can handle potentially millions of users. See Figs. 3, 4, 5, 6 and 7 for screenshots or flow charts of the currently implemented platform.

Plus, the point is once ongoing collaborations materially develop in this way, there are multiplier effects toward achieving the many other social goals of the SDGs that more innately come about only once we set up material sustainability and its dynamics as a priority in all regions worldwide.

4.1 Project Description

For more details about the rubrics for Commodity Ecology, it comprises a set of 130 different material/technological categories. Some categories are: textiles, dyes/colorants (murex, cochineal, synthetic chemicals, derived organic coal-based chemicals), building materials/tool construction, metals, and garbage/garbage disposal. Ecoregions are geographical regions characterized by unique ecological patterns among soil type, flora and fauna, and microclimate. There are 867 ecoregions representing how ecosystem dynamics differ across the globe, and these are very stable unlike political borderlines and state capacities. The detailed building blocks of the CEMVC platform are shown in Fig. 2.

The Commodity Ecology Mobile Virtual Community (CEMVC) platform is aimed at creating a convergence of real/virtual debate and documentation about material sustainability in a particular ecoregion or watershed and how to improve it. Once CEMVC is globally available, the participants in this community can ask three interrelated questions in the ecoregions worldwide:

1. "Do we have enough sustainable choices available in a particular material category yet, in this particular region?", and
2. "Are we choosing well in this category toward sustainability yet, in this particular region?", and
3. "How might we help local consumers, producers and the environment simultaneously by understanding what products or wastes in one commodity category might be more productively used in other categories?".

The goal is hardly 'filling in all 130 categories' in all ecoregions, or competing between ecoregions. The goal is to use the online platform and rubric as a process of inventory taking, to develop an ongoing conversation about democratic risk assessment and business incubation toward what particular categories a region wants to focus upon for the future. The goal is to learn laterally across multiple ecoregions instead of compete. For instance, if one zone has learned how to effectively link various commodity categories in a particular kind of industrial production and waste stream integration into other industries, while choosing better materials, other regions can learn from the best practices of others for how to do this. The Commodity Ecology rubric makes it easy to understand these relationships as well as toggle between

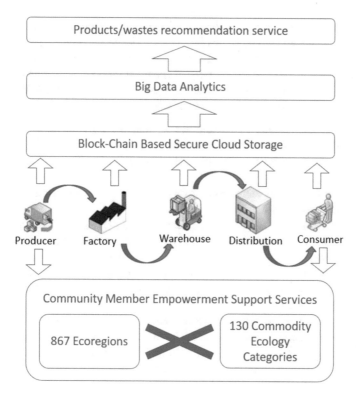

Fig. 2 Building blocks of Commodity Ecology Mobile Virtual Platform

different ecoregions in one's online analysis of the relationships, problems, or priorities at hand in different regions both for how they differ and for what they have in common that might be applied elsewhere. This is not conceived of as a replacement of a globalized economy, though as a useful and sorely required 'ecological check and balance' to maintain our world's cultural diversity and regional biodiversity as intertwined with multiple regional economies nested within such a globalized economy.

4.2 Commodity Ecology Provides Four Kinds of Empowerment Support and Eleven Major Services for "Smart Regions"

The CEMVC provides four types of empowerment support to community members: instrumental support (trading of products/wastes within ecoregions; finding business partners or products), informational support (sharing information about products/wastes and making equally a useful single platform for business/industry news about all sustainable materials in all categories in one place), emotional support (finding friends and networking with like-minded individuals whether globally or in particular real ecoregional areas), and appraisal support (providing recognition for good ideas that build regional sustainable development). The CEMVC provides eleven services to empower members as a 'smart region':

1. Facilitate discussions on commodities' sustainability among people in the same as well as different ecological regions;
2. View, like, share and comment on posts created by other users in their ecoregion or globally;
3. Create posts within their ecoregion or globally by selecting one or multiple commodities categories;
4. Cross reference posts across multiple commodity categories and regions;
5. A search function that filters archived posts based on commodity categories and ecoregions;
6. Enroll the member as a producer or consumer in a particular ecoregion to advertise and find others; you might call Commodity Ecology an "Ecological Facebook" on this issue.
7. Enable producers to list available products and wastes so other producers and consumers can see available products and wastes in their ecoregions to buy them from producers to minimize polluting the region's commons;
8. Provide an ability for producers and consumers to contact each other and share products and wastes;
9. Encourage sustainable market exchanges of materials between producers to consumers (in products) and between producers (as a 'waste trading service' in a region between producers without polluting the environment) thus encouraging profitable, sustainable sharing of products and wastes; you might call Commodity Ecology an "Ecological eBay" for 'smart regions.'
10. Education via gamification built into the platform to inspire innovative youth entrepreneurs to contribute early to a sustainable world in their region and to be recognized by their peers and adults for their ideas or comments; and
11. Invent a useful single platform for business/industry news and trends about all sustainable materials in all categories in one place.

4.3 Seven Functional Requirements of Commodity Ecology

A robust architecture and technical platform are necessary to support member mobility and empowerment needs of CEMVC members. Figure 2 shows the building blocks of CEMVC. Seven main functional requirements of CEMVC are:

1. Community member roles: The platform must support the creation of the following user roles and associated access rights: unregistered users (able to read

available posts on global zones or ecoregional zones), registered users (able to create new posts and comment on posts only in particular ecoregions) and moderator (able to configure commodity ecology categories plus creating posts or deleting posts or misbehaving users; able to create or to remove sub-moderator privileges per each ecoregion upon certain already registered users);

2. Access and Resiliency: The CEMVC platform and its functionalities must be accessible using smart phones, tablets and desktop computers. Interfaces must adapt to device and networking capabilities; capacity of uploads/downloads should ideally be distributed as well perhaps employing architectures using IPFS (Interplanetary File System) protocols and/or allowing for multiple voluntary hosting running on various computers worldwide, similar to how Bitcoin 'mining' is done currently;

3. Languages and time zones: The CEMVC platform must support multiple languages and time zones as this is intended to be a globally available information service;

4. Ease-of-use/usability: The user-interface must be intuitive and easy to use; user role, skills, age, gender and any disability may affect the user-interface requirements;

5. Information access control: The posts linked to one or multiple commodities and/or ecoregions should be available only in those parts; deep levels of encryption protect user private information and passwords from hacking;

6. Facilitate sustainable product/material journey: The platform should have provision to encourage or even to track origin and transfer of products/wastes purchases or even bidding via technologies such as Blockchain technology if the platform is to be used equally to facilitate purchases within or across regions;

7. Data analytics and recommendation services: The platform should increasing use some kind of artificial intelligence (AI) to provide data analytics services for users to analyze effectiveness of CEMVC for sustainable development as well as to recommend products/wastes to producers/consumers based on their preferences, as well as provide an information service that summarizes the category-specific commodity business trends toward sustainability development in one convenient place.

Added gamification would be useful for educating and drawing the youth demographic to the platform as well. For lack of space, these ideas of gamification are left unelaborated in this short article though the ideas for it are very detailed and prepared already (Figs. 3, 4, 5 and 6).

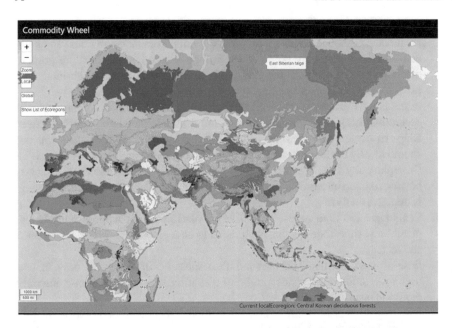

Fig. 3 Map of global ecoregions—clicking on a colored ecoregion brings users to that region's commodity wheel of posts of sustainable ideas in all 130 categories of commodities, in Fig. 4

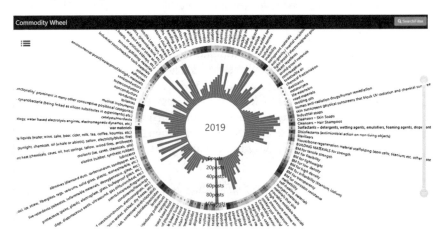

Fig. 4 Commodity Wheel showing scale of posts in that particular ecoregion, per commodity category—next, clicking on a commodity category brings a user to that particular ecoregion/commodity's posts in Fig. 5

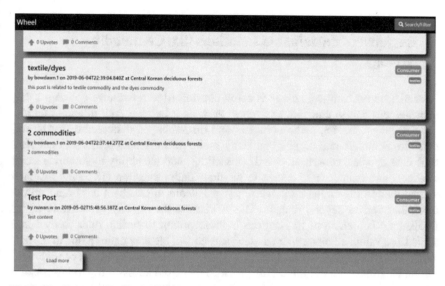

Fig. 5 Feed of posts in a particular commodity/ecoregion combination

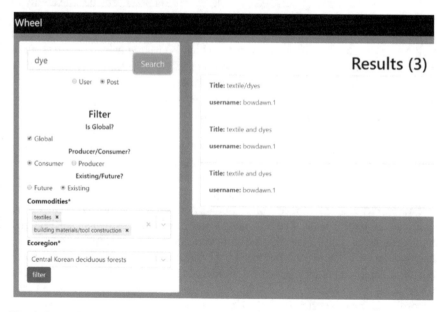

Fig. 6 Search function for archived sustainable ideas and commodity linkages, filtered either by user, post topic, global/'my local ecoregion', consumer/producer, future/existing idea, commodity category, or particular ecoregion

5   Applying Decentralized Ledgers to Manage 'Smart Regions' of Sustainable Relations like Commodity Ecology

Recently, blockchain technology is being considered as a lucrative option for data management due to compelling features such as it being a fault tolerant and distributed ledger, having a chronological and time-stamped data record that is irreversible, auditable and cryptographically sealed in information blocks, near real-time data updates, consensus-based transactions, and the ability to authorize smart contracts and policy based access to facilitate data protection [18]. First invented and applied in the online cryptocurrency of Bitcoin, Blockchain has branched out into other venues beyond finance. This is because the underlying software of the Blockchain is a chain of transactions built according to certain rules formed in a signed and validated block whereas each subsequent block contains a link to the previous one. Blockchains are cost-effective distributed ledgers without a central monitoring clearinghouse required. This provides increased accessibility to information by connecting stakeholders directly without requirements for such third-party brokers. Compared to existing solutions, Blockchain technology provides the ability to safely and transparently make changes from an unlimited number of sources that can be geographically distributed. Moreover, use of blockchain technology ensures that each participant has the latest data version which is ensured using consensus algorithms in which the system is decentralized and there is confidence in the validity of each transaction. In the CEMVC, we envision the application of blockchain technology to facilitate the sustainable product/material journey by tracking origin and transfer of products/wastes purchases and supporting transparent bidding process. The product/material journey as envisioned in CEMVC is similar to product/material journey in supply chain management—where the material, information and services flow from a producer to consumer. Figure 7 shows application of blockchain technology in CEMVC for facilitating product/waste journey and for enabling product/waste recommendations to the consumers (Fig. 7).

The applications of blockchain technology in supply chain management are described in several texts. The advantages and disadvantages of using RFID and blockchain technology in building the agri-food supply chain traceability system are discussed in [19]. The system proposed in [19] realizes the traceability in the entire agri-food supply chain, by gathering, transferring and sharing the authentic data of agri-food in production, processing, warehousing, distribution, and selling links. According to the discussion in [20] Blockchain is suitable for financial transactions where no physical goods change hands (such as financial instruments, derivatives, and bidding processes) and also for supply chain management, assisting in the delivery of unadulterated, source, process and transit-verifiable goods and commodities. In the area of supply chain management, sustainability is defined a triple-bottom-line concept that includes a balance of environmental, social, and business dimensions when managing the supply chain. There is an increasing momentum towards building sustainable solutions using blockchain technology. A recent article [21] discusses

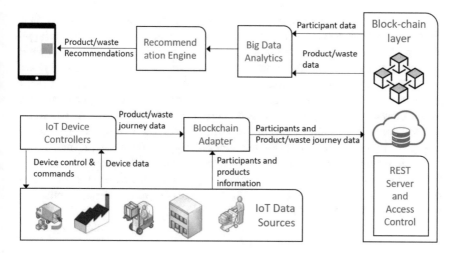

Fig. 7 A block-diagram showing application of blockchain technology in CEMVC

the potential of blockchain technology to contribute to sustainability of social, environmental and recycling supply chains. This article [21] also points out that future research could be focused on using the environmental and social dimensions of sustainability such as the U.N.'s Sustainable Development Goals (SDGs) to evaluate Blockchain-enabled supply chain effectiveness.

5.1 Public Versus Private Block-Chain Network as an Implementation Choice

When using Blockchain to facilitate the social trust that moves product/waste material journeys from the producer to consumer, it is required to consider the choice of Blockchains between public and private versions. A public Blockchain network is completely open for joining and participation in the network, implying little to no privacy for transactions. A public Blockchain needs a substantial amount of computational power to solve a proof-of-work problem and maintain a distributed ledger at a large scale. Hence, public Blockchains are not suitable for our purpose. Private Blockchains also called as 'Permissioned Blockchains' grant specific rights and restrictions to participants in the network. Participants need to obtain an invitation or permission to join. The access control mechanism varies: existing participants may decide future entrants; a regulatory authority could issue licenses for participation; or a consortium could make the decisions instead. After joining the network, the entity plays a role in maintaining the Blockchain in a decentralized manner [22]. Because of these security and privacy features, private Blockchains are more suitable for facilitating product/waste material journey.

There exist other frameworks of distributed ledger technology such as Ethereum, HydraChain, and Hyperledger Fabric that allow the development of Blockchain applications [23]. Ethereum is a decentralized platform that runs smart contracts which are applications that run exactly as programmed without any possibility of downtime, censorship, fraud or third-party interference. Ethereum is primarily a public Blockchain, however HydraChain is an extension of the Ethereum platform which adds support for creating permissioned Blockchains. The Hyperledger Fabric is a permissioned Blockchain framework implementation, and one of the Hyperledger projects is hosted by The Linux Foundation. The Hyperledger Fabric leverages container technology to host smart contracts called 'chaincode' that comprise the application logic of the system. Both, the Hyperledger Fabric and Ethereum allow smart contract services to be exposed as REST APIs, thus allowing external applications to use the Blockchain smart contracts.

6 Conclusion: Commodity Ecology as an Ideal Ecological Design Process

Van der Ryn and Cowan argued in *Ecological Design* (1996) [24] that one can judge and implement a good ecological design as having five key principles. This section will argue that a hard to define 'sustainable smart city' is far easier to define as a 'sustainable smart region'–given the discussion above about media regimes and given the kind of social-technical world of communication we live in now. This 'smart regional' world we live in now of deeply saturated mobile ICT networks can be contrasted to the 'smart city' world we lived in in 1996 with very geographically limited and expensive uses of such networks.

The first of the five principles of ecological design is the "details of the place." Thus, an actual ecological design is clearly better implemented as an *urban and regional development* instead of just as an urban development per se because the context is everything. "The first principle grounds the design in the details of place. In the words of Wendell Berry, we need to ask, 'What is here? What will nature permit us to do here? What will nature help us to do here?' [24] Commodity Ecology certainly fulfills that criterion as it embraces a wider and more participative 'smart region' instead of only framing itself as an urban-based improvement of a 'smart city' where the latter is without awareness or feedback from the particular ecological situations in its hinterlands per se.

The second principle of ecological design "provides criteria for evaluating the ecological impacts of a given design" [24]. Commodity Ecology does this as well, given it is an ongoing permanent self-evaluation platform and social process that evaluates all material choices for how well they fit within a particular regional context, and how well waste handling happens between them all.

The third principle of ecological design "suggests that…impacts can be minimized by working in partnership with nature" [24]. The idea of building rubrics of

Commodity Ecology and its model of sustainability on recognition of actual regional ecological variety and respecting that as the basis of the deliberation process around Commodity Ecology fulfills that criteria.

The fourth principle of ecological design "implies that ecological design is the work not just of experts, but of entire communities" [24]. If anything is clear at this point, Commodity Ecology is an ongoing collective democratic risk assessment about all materials in a region in which the entire community is foreseen as participating in such a platform for all time—or for as long as the media regime of saturated mobile phone networks last. This kind of media regime is likely to continue because in retrospect, the history of media seldom goes 'backwards' toward reversion to a previous regime. When humans have a fresh communication technology, it becomes part of their wider layered media ecology instead of opposition to it ever amounting to much besides a fringe rejection of it. As long as we live in mobile ICT networks, the entire community of youth to elderly, environmental underclasses [25] to enfranchised elites, men to women, rich urban areas to poor rural areas, central clean places to toxic borderlands, places of consumption to places of garbage disposal, is equally enfranchised in the discussion.

The fifth principle of ecological design "tells us that effective design transforms awareness by providing ongoing possibilities for learning and participation" [24]. This is a principal that can clearly come from such a Friere'ian 'conscientization' [26, 27] that daily would be an ongoing education about oneself and one's ecoregion if one was to participate in such a platform as Commodity Ecology.

All five principles of ecological design are a checklist of what a truly sustainable and ecologically sound development would be. It is doubtful that any 'smart city' in the world could ever claim all five levels of ecological design because conceptually it is rooted in assumptions about a delimited space and a limited ICT coverage of a previous "media/society" relationship that is rapidly being eroded and outmoded toward a civilization based on deeply equal urban and rural mobile ICT networks of participation and synchronous/asynchronous deliberation. This is pressuring all social institutions to adapt. Conceptions of 'smart city' should adapt into 'smart regions' as a consequence. Institutions and developmental models that fail to develop such wide deliberation may be readily replaced—hardly by violence and more by simply ignoring them for other kinds of integration on online platforms that are more participative.

In the introductory quote, it was argued in 1996 that sustainability can be "solved only if industrial designers talk to biogeochemists, sanitation engineers to wetland biologists, architects to physicists, and farmers to ecologists" [28]. Upon reflection about ICT-saturated 'smart regions' now, the quote should be updated and rewritten for 2020 to include how online platforms like Commodity Ecology are required for such deliberations of ecological design to facilitate such discussions. Commodity Ecology is one such 'smart regional' platform that brings everyone mentioned above into the deliberation while adding in all producers, all consumers, and all citizens as well.

Sharing a common future vision is important. The best way to develop a better future is to invent it. Such an ecological design would be to get people, professions, and industries to talk people they rarely talk to in their daily life, in an ongoing democratic risk assessment with their other producers and with their consumers in a 'smart region.' Therefore, a blockchain-based virtual community like Commodity Ecology simplifies this intercommunication. Commodity Ecology operating in every ecoregion of the world due how every region is becoming a 'smart region' is this common vision of the future.

References

 1. Innis HA (1950/1986) Godfrey D (ed) Empire and communications. Press Porcépic Ltd., Victoria (with the assistance of the Canada Council)
 2. Castells M (ed) (2004) The network society: a cross-cultural perspective. Edward Elgar, Cheltenham
 3. Castells M (2004) Informationalism, networks, and the network society: a theoretical blueprint. In: Castells M (ed) The network society: a cross-cultural perspective. Edward Elgar, Cheltenham, p 6
 4. Singh M, Singh A, Kim S (2018) Blockchain: a game changer for securing IoT data. In: 2018 IEEE 4th World Forum on Internet of Things (WF-IoT). IEEE, pp 51–55
 5. We Are Social/Hootsuite (2019) Digital 2019: global internet use accelerates. https://wearesocial.com/blog/2019/01/digital-2019-global-internet-use-accelerates. Accessed 16 Oct 2019
 6. Mancini P (2014) How to upgrade democracy for the internet era. Ted.com https://www.ted.com/talks/pia_mancini_how_to_upgrade_democracy_for_the_internet_era#t-148748. Accessed 16 Oct 2019
 7. Höjer M, Wangel J (2014) Smart sustainable cities: definition and challenges. In: ICT innovations for sustainability. Springer, Berlin pp 333–349
 8. We Are Social/Hootsuite (2019) Global social media users pass 3.5 Billion. https://wearesocial.com/blog/2019/07/global-social-media-users-pass-3-5-billion. Accessed 16 Oct 2019
 9. Pasquale F (2015) The black box society: the secret algorithms that control money and information. Harvard University Press, Cambridge
10. Standing G (2010) The precariat: the new dangerous class. Bloomsbury Academic, London
11. Madhusudan S, Kimb S (2019) Blockchain technology for decentralized autonomous organizations. Role of Blockchain Technology in IoT Applications vol 115, p 115
12. United Nations Academic Impact (UNAI) (2018) #SDGsinAcademia: Goal 12. https://academicimpact.un.org/content/sdgsinacademia-goal-12; https://academicimpact.un.org/content/commodity-ecology-initiative-facilitate-sustainable-development. Accessed 16 Oct 2019
13. Whitaker M (2009) Commodity ecology: the blog https://commodityecology.blogspot.com. Accessed 16 Oct 2019 or https://comwheel.azurewebsites.net
14. Trinade EP, Hinnig MPF, Moreira da Costa E, Marques JS, Bastos RC, Yigitcanlar T (2017) Sustainable development of smart cities: a systematic review of the literature. J Open Innov Technol Mark Compl 3(3):11. https://doi.org/10.1186/s40852-017-0063-2
15. Oh M, Larson JF (2019) Chapter 4: Korea's smart cities and urban information culture. In: Digital development in Korea: lessons for a sustainable world, 2nd edn. Routledge, London (Routledge Advances in Korean Studies)
16. See Updated Mobile Coverage for States. https://www.t-mobile.com/coverage/coverage-map; https://www.opensignal.com/networks; http://maps.mobileworldlive.com/
17. Park G (2019) Rethinking mobile diffusion: explanatory analysis of factors affecting mobile diffusion in african countries in the sub-saharan region. Part A [Master's] Paper, Department of Technology and Society, State University of New York, Korea. Available: park.gayoung@sunykorea.ac.kr (Dissertator); mark.whitaker@sunykorea.ac.kr (Adviser)

18. Shafagh H et al (2017) Towards blockchain-based auditable storage and sharing of IoT data. In: Proceedings of the 2017 on cloud computing security workshop. ACM
19. Tian F (2016, June) An agri-food supply chain traceability system for China based on RFID & blockchain technology. In 2016 13th international conference on service systems and service management (ICSSSM). IEEE, pp 1–6
20. Apte S, Petrovsky N (2016) Will blockchain technology revolutionize excipient supply chain management? J Excip Food Chem 7(3):910
21. Saberi S, Kouhizadeh M, Sarkis J, Shen L (2019) Blockchain technology and its relationships to sustainable supply chain management. Int J Prod Res 57(7):2117–2135
22. Jayachandran P (2017) The difference between public and private blockchain. Blockchain pulse: IBM Blockchain Blog
23. Nagpal R (2017) 17 blockchain platforms—a brief introduction. Medium Blockchain Blog
24. Van der Ryn S, Cowan S(1996) Ecological design, 2nd edn. Island Press, Washington
25. Pellow DN (2008) Chapter 7: Environmental racism: inequality in a toxic world. In: Romero M, Margolis E (eds) The Blackwell companion to social inequalities. Wiley-Blackwell, Hoboken, pp 147–164
26. Garavan M (2010) Opening up Paulo Freire's Pedagogy of the Oppressed. In: Dukelow F, O'Donovan O (eds) Mobilising classics: reading radical writing in Ireland. Manchester University Press, Manchester
27. Shih Y-H (2018) Some critical thinking on Paulo Freire's critical pedagogy and its educational implications. Int Educ Stud 11(9):64–70
28. Van der Ryn S, Cowan S(1996) Ecological design, 2nd edn. Island Press, Washington, pp ix–x

Toward Multiple Layered Blockchain Structure for Tracking of Private Contents and Right to Be Forgotten

Min-gyu Han and Dae-Ki Kang

Abstract We propose a novel Blockchain architecture for providing services with insertion and deletion features without compromising decentralization and integrity. In our proposed architecture, we construct three stage multiple layered blockchain. In previous blockchain architectures, it is difficult to modify the contents and contents-index because they are in one public blockchain. In our architecture, extra blockchain manages the link between contents and content-index to resolve the difficulty in modifying the contents. We present various use cases of our multiple layered blockchain architecture to demonstrate its effectiveness.

Keywords Next generation blockchain platform · Multiple layered structure · Private contents · Right to be forgotten

1 Introduction

"Right to be forgotten" has become very serious problems in our society. In particular, because of the inherent difficulty in deleting data and update history of blockchain based services, it is impossible to modify and delete the contents and private information already uploaded. Moreover, this impossibility of contents removal basically violates information assurance laws such as general data protection regulation (GDPR) law in Europe [1].

In terms of technology, there have been methods, such as hardfork, for deleting and modifying blockchains [2]. However, those previous methods have destroyed the relation with legacy data, thus exhibit problems in continuity and stability of services and cannot support frequent modification of data.

M. Han
ICT Convergence Program, Hansung University, Seoul, South Korea
e-mail: andyhan@hansung.ac.kr

D.-K. Kang (✉)
Department of Computer Engineering, Dongseo University, Busan, South Korea
e-mail: dkkang@dongseo.ac.kr

© Springer Nature Singapore Pte Ltd. 2020
D. Singh and N. S. Rajput (eds.), *Blockchain Technology for Smart Cities*,
Blockchain Technologies, https://doi.org/10.1007/978-981-15-2205-5_5

1.1 Background

In Korea, the "right to be forgotten" is clearly preteded from Personal Information Protection Act. From Article 21 item (i) of Personal Information Protection Act of South Korea [3], it is clearly written that *'when personal information becomes unnecessary as its holding period expires, its management purpose is achieved and by any other ground, a personal information manager shall destroy the personal information without delay: Provided, That this shall not apply where the personal information must be preserved pursuant to any other Act or subordinate statute.'*

Also, Korea Communications Commission (KCC) have hold several discussions with industrial experts on the right to be forgotten [4]. In 2016, KCC enacted 'Exclusion Request Access Rights of Internet Self-Post Guidelines' to guarantee Internet users' right to be forgotten [5].

Many private blockchain tech companies simply make the excluded blockchain unsearchable. For example, Korea Telecom (KT) developed 'KT data chain architecture' [6]. KT data chain is a blockchain that records digital data and has a deletion feature.

There have been many worldwide activities for protecting the "right to be forgotten".

In May 2014, European Court of Justice (ECJ) certified the right to be forgotten in General Data Protection Regulation (GDPR) of European Union (EU) [1]. Note that this includes the right to remove the data description including its pointers from the search results.

Russia enacted the law to request removal of personal information which can be retrieved from Internet [7].

As for Google, users can submit removal request to protect their personal information using Google's removal tool [8]. They can fill out the removal form to specify what information they want to remove from the search results and why they want to remove it.

Table 1 shows the technical trends for protecting private information.

According to Google Transparency Report [9], the number of requests to delist content under European privacy law is 843,411 in September 2019, and the number of URLs requested to be delisted is 3,325,311. As you see in Fig. 1, there has been a drastic increase in the number of requests.

1.2 Summary of Development

Development of the final product: Against this background, to solve the problem of the right to be forgotten over blockchain, we propose a solution for tracking and modification/deletion of private contents based on multiple layered blockchain structure. The summary of our development is as follows:

Table 1 Technical trends for protecting private information

Method	Description	Drawbacks
Off chain storage	Private information is saved outside the block	No decentralization
		Vulnerable to hacking and manipulation
		Access right to private information is on-chain
Blacklist	Use of encrypted key to access private information	Removing the key makes information unaccessible, but unable to track its history
Hard Fork	Implement hard fork for private information to remove	Impractical

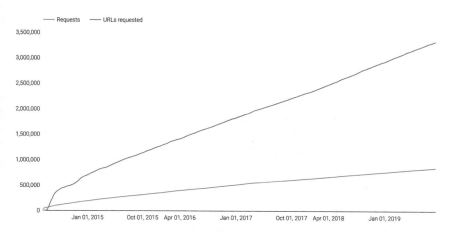

Fig. 1 Requests to delist content under European privacy law

- Solution for tracking and modification/deletion of private contents based on multiple layered blockchain structure
  - Multiple layered blockchain basic service
  - Multiple layered blockchain based user management service
  - Multiple layered blockchain based private contents management service.
- Application interface for supporting multiple layered blockchain structure
  - User function interface
  - Blockchain interface
  - Private contents interface.

Development Goal in Incremental Release Plan: The development is performed in terms of multiple releases. We summarize two step releases of the system as follows:

- The first release is on the development of prototypes for solving real world problems

  - Requirement gathering and service scenario construction
  - Basic level design of the system
  - Multiple layered blockchain based service prototype for tracking and modification/deletion of private contents private contents management service.

- The second release is on the development of commercial solution for solving real world problems

  - Platform construction of main functions
  - Blockchain social media service and interface with other applications
  - Test of commercial solutions, based on service scenarios.

2 Related Work

2.1 Commercial Products

In terms of commercial developments, there are applications that support "the right to be forgotten". For example, Snapchat[1] removes users' contents (Snaps and Chats) in friends' smartphones after 24 hours. Also, following GDPR's regulations, Snapchat preserve less data for youths than adults.

And, in Korea, Eggshot[2] interacts with KakaoTalk application and automatically removes comments, messages and pictures after one day.

MARACHAIN[3] is a blockchain architecture to save personal information in a distributed fashion using InterPlanetary File System (IPFS). Users' private information are encrypted when they are transmitted.

2.2 Intellectual Properties

With respect to intellectual properties, there are several patents filed for implementing "the right to be forgotten". Some of them are listed as follows:

- "The Digital Aging System and The Management Method" (Patent No. KR 1012583870000): in this invention, they apply the idea of lifespan to digital data. In other words, this invention considers data as an living organism with aging, and manages the history of data usage. The aging mechanism in the data reflects

---

[1]https://www.snapchat.com/.
[2]http://eggshot.me/.
[3]https://www.marachain.eu/.

utilization of the data, so that the user can easily decide whether to preserve or eliminate the data.

- "File Uploader, Digital Aging Agent, Management System and Method thereof" (Patent No. KR 1020150103649): in this invention, when a user uploads files to the content upload server, the server sends the user a file uploader program with destruction time setting so that the user can upload their aging contents.

2.3 Standardization

Our proposed architecture is relevant to the following standardization activities:

- Domestic standardization

  - Intelligent Contents Standardization Forum is working on Intelligent Blockchain Standardization Task.
  - Telecommunications Technology Association (TTA), Finance Security Standardization Association, Personal Information Protection Standard Forum, and Blockchain Standardization Forum are discussing Blockchain based Personal Information Protection Guidelines.
  - Blockchain Standardization Forum is preparing Blockchain based Electronic Document Exchange.

- International standardization

  - International Organization for Standardization (ISO) is discussing Blockchain based Personal Information Protection Service Architecture and Requirements.
  - ISO is also working for Use cases of Blockchain and Distributed ledger.
  - OneM2M is discussing Trust management in oneM2M Blockchain and Use cases for one M2M.

3   Development Strategy

Our development strategy is summarized as follows:

- Implementation and industrialization through cooperative work between industry sector and academy sector.

  - Patent filing for blockchain management and contents management.

- Regular/irregular workshop and meeting for cooperative development and verification/validation
- Cooperation with open software foundation and consulting from external experts
- Multiple layered blockchain based service prototype for tracking and modification/deletion of private contents private contents management service.

4 Our Method

4.1 Outline

The goal of our proposed method is to provide services that can modify and delete data based on blockchain with preserving decentralization and integrity. Note that decentralization and integrity are the important features of blockchain.

Our solution is by constructing three stage multiple layered blockchain (Fig. 2). Our three stage multiple layered blockchain introduces extra blockchain to manage the link between contents and content-index. The extra blockchain resolves the difficulty in modifying the contents and contents-index when they are in one public blockchain, which is common in previous blockchain architectures.

Our multiple layered blockchain architecture is used for private contents management (Fig. 3). The architecture is composed of public service blockchain (PSB), private management blockchain (RMB), and public contents blockchain (PCB). PSB is in charge of storing information used for the contents service. RMB is responsible for storing a link information between the service and encrypted contents. Finally, PCB is for storing the encrypted contents and contents information (such as P2P storage URI). We will explain Fig. 3 more detail in the later section (Sect. 4.3) when we explain use cases for our proposed architecture.

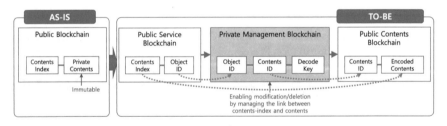

Fig. 2 Multiple layered blockchain structure for tracking private contents

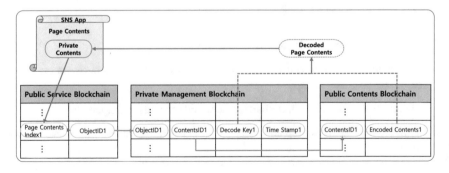

Fig. 3 Private contents creation using multiple layered blockchain

The advantages of the proposed architecture and method are as follows:

- Contents can be modified and deleted by adding data to multiple layered blockchain.
- Through multiple layered blockchain architecture, private contents can be tracked, modified, and deleted in batch fashion.
- If it is legally required, permitted official can check the old contents before modification or deletion.

Our development is summarized as follows:

- Solution for tracking and modification/deletion of private contents based on multiple layered blockchain structure

    - Multiple layered blockchain basic service
    - Multiple layered blockchain based user management service
    - Multiple layered blockchain based private contents management service.

- Application interface for supporting multiple layered blockchain structure

    - User function interface
    - Blockchain interface
    - Private contents interface.

The goals of our releases are projected as follows:

- Blockchain transaction processing speed (See Table 2)
- Contents server's response time: one or two seconds in multiple layered blockchain environment
- Block processing related indices in blockchain (See Table 3).

The key technologies for our proposed architecture are contents management in multiple layered Blockchain and personal contents tracking, modification and deletion in Blockchain. Our architecture can be applied to public blockchain service which deals with a large amount of personal information, and it can be used in blockchain based social media service and web sites where personal contents are frequently exposed.

Table 2 Blockchain transaction speed

	Unit	Before	First release	Second release
Transaction processing speed	sec	7	7	20
Transaction processing speed with multiple blockchain	sec	3	3	6

Table 3 Block record related indices

	Unit	Before (KB)	First release (KB)	Second release (KB)
Block record size	sec	7	7	20
Block creation time	sec	15	14	12

4.2 Architectural Detail

The architecture of our proposal is depicted in Fig. 4.

As is described in Fig. 4, the private contents blockchain system based on multiple layered blockchain has three different types of blockchains. As aforementioned in Sect. 4.1 , they are (1) "public service blockchain" (PSB) which is directly related with service, (2) "private management blockchain" (RMB) which manages links between service and contents, (3) and "public contents blockchain" (PCB) which is in charge of encrypted contents.

Blockchain nodes of public service blockchain and public contents blockchain are constructed by service users, and blockchain nodes of private management blockchain are constructed by service managing company.

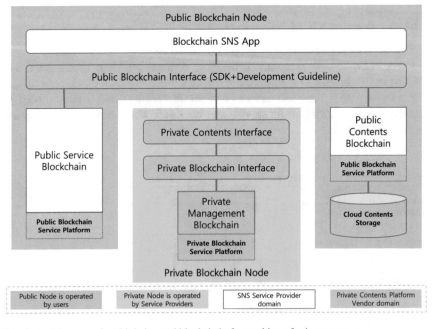

Fig. 4 Architecture of multiple layered blockchain for tracking of private contents

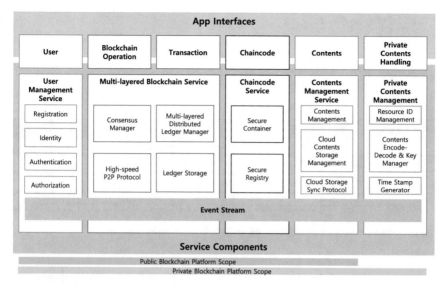

Fig. 5 App Interfaces of multiple layered blockchain for tracking of private contents

Our multiple layered blockchain solution is composed of "service components" and "app interfaces" (Fig. 5). Service components are for providing services using blockchain, and app interfaces are for supporting service app development.

Service components include user management service, multiple layered blockchain service, chaincode service, contents management service, and private contents management service. App interfaces provide access to each service component.

4.3 Service Use Cases

We explain use cases to explain the fundamental operations our proposed multiple layered blockchain in terms of private contents management.

Private Contents Creation: Figure 3 shows the use case of private contents creation in multiple layered blockchain. Use case scenario for this is as follows:

1. Issue "ObjectID1" is matched with the Private Contents ("Private Contents Index1"). The Private Contents could be a word, a sentence, a web page, an image, a video file and extra.
2. Add the "Private Contents Index-ObjectID1" on the TOP (or the LAST) BLOCK of "Public Service Blockchain (PSB)".
3. Encode the Private Contents and create "Decode Key1", issue the "ContentsID1" for the encoded contents and add the "ContentsID1-Encoded Contents1" on the TOP (or the LAST) BLOCK of "Public Contents Blockchain (PCB)".

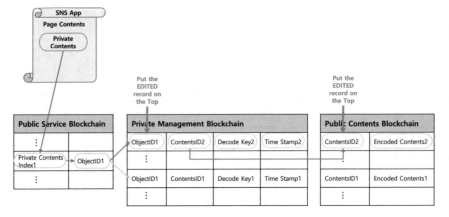

Fig. 6 Private contents modification using multiple layered blockchain

4. Issue the "Time Stamp1" at the moment and add the "ObjectID1-ContentsID1-Decode Key1-Time Stamp1" on the TOP (or the LAST) BLOCK of "Private Management Blockchain (RMB)".

Private Contents Modification: Figure 6 shows the use case of private contents modification in multiple layered blockchain.

Use case scenario for this use case (Fig. 6) is as follows:

1. Encode the new Private Contents and create "Decode Key2", issue the "ContentsID2" for the encoded contents and add the "ContentsID2-Encoded Contents2" on the TOP (or the LAST) BLOCK of "Public Contents Blockchain (PCB)"
2. Issue the "Time Stamp2" at the moment and add the "ObjectID1-ContentsID2-Decode Key2-Time Stamp2" on the TOP (or the LAST) BLOCK of "Private Management Blockchain (RMB)".

Private Contents Deletion: Figure 7 shows the use case of private contents deletion in multiple layered blockchain.

The use case scenario for private contents deletion is as follows:

1. Get ObjectID1 of Private Contents to edit from the Page Contents
2. Encode the "NULL Private Contents" and create "Decode Key2", issue the "ContentsID2" for the encoded contents or simply select from the "pre-registered NULL Contents" (It could be a default text as "Deleted" or an image indicating deleted contents) and add the "ContentsID2-Encoded Contents2" on the TOP (or the LAST) BLOCK of "Public Contents Blockchain (PCB)"
3. Issue the "Time Stamp2" at the moment and add the "ObjectID1-ContentsID2-Decode Key2-Time Stamp2" on the TOP (or the LAST) BLOCK of "Private Management Blockchain (RMB)" or, simply notify "NULL" or "VOID" on the place of ContentsID2 and add the "ObjectID1-NULL(VOID)-Decode Key2-Time

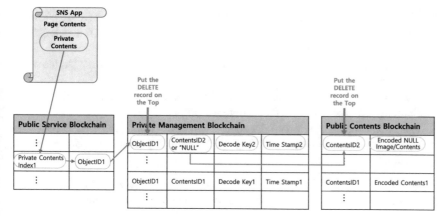

Fig. 7 Private contents deletion using multiple layered blockchain

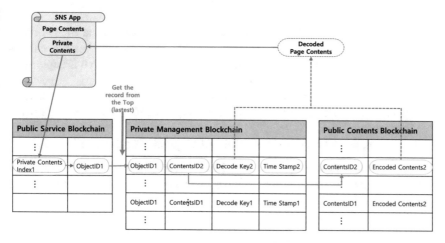

Fig. 8 Displaying modified private contents using multiple layered blockchain

Stamp2" on the TOP (or the LAST) BLOCK of "Private Management Blockchain (RMB)".

Private Contents Display: Figure 8 shows the use case when the multiple layered blockchain displays modified private contents.

Detailed steps for this use case is explained in the following scenario:

1. Social media app (SNS App in Fig. 8) requests to get the "ObjectID1" with Private Contents Index in its contents page from the Public Service Blockchain(PSB)
2. The social media app requests the Private Contents with the "ObjectID1" to the social media service provider

3. The service provider reads the "Private Management Blockchain (RMB) from the Top (latest), finds the first "ObjectID1" record, and gets "ObjectID1-ContentsID2-Decode Key2-Time Stamp2".
4. The service provider replies the "ContentsID2" and "Decode Key2" to the social media App.
5. The social media app requests to get the "Encoded Contents" to the "Public Contents Blockchain" with the "ContentsID2"
6. The social media app gets "Encoded Contents2" from the PCB, decodes the encoded contents with the "Decode Key2" and display on the place of the "Private Contents".

Private Contents Distribution: Figure 9 shows the use case when the multiple layered blockchain distributes private contents. This distribution include 'share' and 'follow' operation in social media.

Use case scenario for distributing modified private contents (Fig. 9) is shown below:

1. If social media user makes distribution of Private Contents via Share/Follow which is popular method of social media service, the social media service (SNS App in Fig. 9) issues another "Private Contents Index2" because it is duplicated, and issues "ObjectID2" for the same reason.
2. Issue the "Time Stamp2" at the moment and add the "ObjectID2-ContentsID1-Decode Key1-Time Stamp2" on the TOP (or the LAST) BLOCK of "Private Management Blockchain (RMB)"
3. As a result, the "ObjectID1" and "ObjectID2" point "ContentsID1" together.

Private Contents Tracking: Here, we discuss the way to track modifications of a particular contents in multiple layered blockchain. Figure 10 depicts use case for this private content tracking.

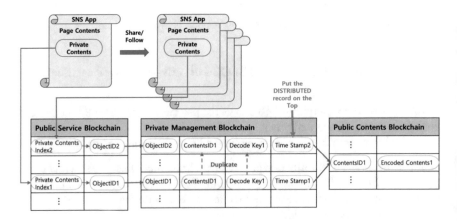

Fig. 9 Distributing modified private contents using multiple layered blockchain

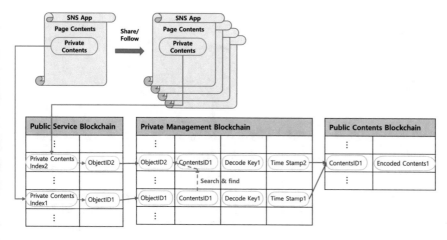

Fig. 10 Tracking modified private contents using multiple layered blockchain

The use case scenario for Fig. 10 is as follows:

1. If social media user wants to find a Private Contents at any reason, the user points the Private Contents of any Page.
2. Social media service app could get the ObjectID ("ObjectID1" as an example) from the PSB and requests to RMB with the ObjectID ("ObjectID1")
3. RMB service gets the ContentsID ("ContentsID1" at this time), and using this, RMB could find the list of ObjectsIDs ("ObjectsID1" and "ObjectID2"). RMB responses the list of ObjectsIDs to social media app, user could recognize the list of pages which have the Private Contents.

Temporal Tracking: We describe how we can track private contents in temporal domain (Fig. 11). In other words, we show how to search old contents in our proposed multiple layered blockchain architecture.

The following is a use case detail of temporal tracking for Fig. 11.

1. If an authorized social media user (e.g. Police departments) wants to see a Private Contents which has been deleted, the social media app (SNS App in Fig. 11) requests the Privates Contents to RMB service with "ObjectID1" and date & time ("Time Stamp1" for example).
2. RMB service finds the "ContentsID1" and replies the "ContentsID1" and "Decode Key1" to the social media app.
3. The the social media app requests to get the "Encoded Contents1" to the "Public Contents Blockchain" with the "ContentsID1"
4. The the social media app gets "Encoded Contents1" from the PCB, decodes the encoded contents with the "Decode Key1" and display on the place of the "Private Contents".

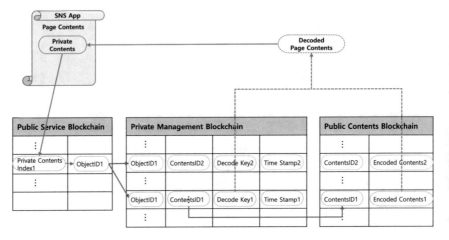

Fig. 11 Temporal tracking modified private contents using multiple layered blockchain

4.4 Development Detail in Incremental Releases

First release: In the first release, we aim to design and experiment a prototype system for multiple layered blockchain based private contents modification and deletion. The prototype includes contents management, user identity, private contents management, and user registration.

Detailed development plan of contents management and user identity usually includes identity in user management, consensus management in blockchain service, multiple distributed ledger management, ledger storage management, and contents management. Please refer to Fig. 12 for more detailed information about the ledger structure of multiple layer blockchain. Detailed development plan of private contents management usually includes resource ID management, encode/decode of contents, and key management. Detailed development plan of user registration includes components for user registration component, user access privilege authentication and grants.

Second release: The second release mainly focuses on commercialization of our solution. The commercial product will include private contents modification and tracking. The goal of the commercial level system is to complete the development of main app interfaces, chaincode and contents management. Main app interfaces contains private contents app interfaces and user function app interfaces. In addition to development of main app interfaces, we need to implement high-speed P2P blockchain synchronization, protocol, chaincode security container, secure registry, cloud contents storage management, and cloud storage synchronization protocol.

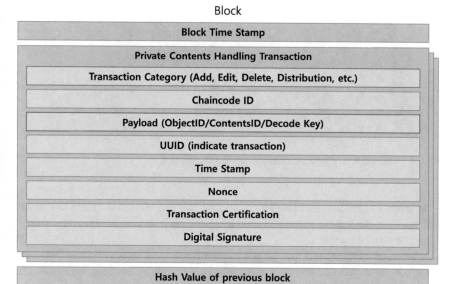

Fig. 12 Ledger structure of multiple layered blockchain

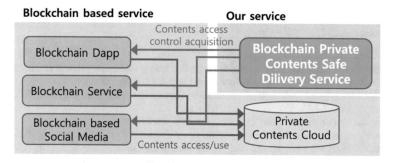

Fig. 13 Blockchain based private contents safe service ASP

5 Future Plan and Expected Effects

Our proposed product can be further used in the following business:

- Blockchain based private contents utilization services (Fig. 13)
- Safe supply of blockchain based solutions for private contents and its consulting service (Fig. 14).

From our product and its business, we can envision the following results:

- In terms of technology, with blockchain based diverse Dapp's and services, we can utilize private contents in a completely destructible, modifiable and stable form.

Blockchain based Service Company

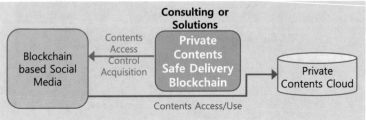

Fig. 14 Blockchain based private contents safe service solution

– In terms of business and industry, since we can use private contents in a stable form with blockchain based services, it will activate service development market and its related markets.
– In terms of social aspects, our proposal can solve the problem of "right to be forgotten", because private contents can become completely destructible and modifiable in our blockchain based service.

References

1. Mantelero A (2013) The EU proposal for a general data protection regulation and the roots of the 'right to be forgotten'. Comput. Law Secur. Rev. 29(3):229–235
2. Lin IC, Liao TC (2017) A survey of blockchain security issues and challenges. Int. J. Network Secur. 19(5):653–659
3. South Korea: Personal Information Protection Act (Mar 2011)
4. Ministry of the Interior and Safety: Guidance on "General Data Protection Regulation" for Korean Enterprises. Tech. rep, Ministry of the Interior and Safety, South Korea (Apr 2017)
5. Ministry of the Interior and Safety (2016) Exclusion Request Access Rights of Internet Self-Post Guidelines. Tech. rep, Ministry of the Interior and Safety, South Korea
6. Suberg W (2019) Korean Telecom Giant KT to Launch Blockchain-as-a-Service Platform in March: Report. Cointelegraph (Feb 2019). https://cointelegraph.com/news/korean-telecom-giant-kt-to-launch-blockchain-as-a-service-platform-in-march-report
7. Freedom on the Net 2018—Russia. Tech. rep., Freedom House (Nov 2018). https://www.refworld.org/docid/5be16afe21.html
8. Remove your personal information from Google (2019). https://support.google.com/websearch/troubleshooter/3111061?hl=en
9. Google Transparency Report: Requests for user information. Tech. Rep. 2019, Google (2019). https://transparencyreport.google.com/?hl=en

Smart City Transportation Technologies: Automatic No-Helmet Penalizing System

Ashutosh Agrahari and Dhananjay Singh

Abstract With the concept of Smart City, the concept of a Smart Traffic Management System comes up automatically. Traffic chaos has increased a lot nowadays. There are now more vehicles than there are men on road. Due to this boom in urbanization and the number of vehicles on road, the problem to keep a check on the riders has also become very difficult for the policemen. People take traffic safety measures for granted and do not take them seriously. As a result, a lot of accidents happen heavily due to not wearing helmet on motorcycles, and not following the safety measures prescribed by the traffic police department of the nation. To overcome this challenge, this solution will act as a helping hand for the policemen in controlling the traffic. The solution will detect the riders for the helmet, and help the policemen to get an actual glimpse of the helmet-wearing status in the city and better control the traffic accidents and enforce the traffic rules. If the rider is found to be not wearing a helmet then his or her number plate will be scanned and stored in the fine database, from where a fine will be generated based on the vehicle registration number that has been captured by the surveillance camera.

Keywords ANPR · Automatic fine generation · Helmet detection · Smart city · Traffic rules · Road accidents

A. Agrahari
Department of Computer Science and Engineering, Amity University, Lucknow, India
e-mail: ashutoshagrahari@acm.org

A. Agrahari · D. Singh (✉)
VESTELLA Lab Inc., Seoul, South Korea
e-mail: dan.usn@ieee.org

D. Singh
Department of Electronics Engineering, Hankuk University of Foreign Studies, Seoul, South Korea

© Springer Nature Singapore Pte Ltd. 2020
D. Singh and N. S. Rajput (eds.), *Blockchain Technology for Smart Cities*,
Blockchain Technologies, https://doi.org/10.1007/978-981-15-2205-5_6

1 Introduction

The term Smart City encircles the people dwelling inside it and, the technologies and devices they use. Everyone wants to live in an urban setting and smart cities are a sort of technological trend moving the people towards more urban settings and lifestyle. With increasing standard of living, the demand for materialistic desires is also increasing with rapid pace. Everyone wants to own at least the basic luxury which includes own house, car, fridge, television, air conditioner, and many more that a man requires to live a so-called satisfying life. Smart cities are connected mostly with mobile devices whether it be a cell phone or a vehicle. This paper mainly focuses on motorcycle related effects in a smart city setting.

With increasing traffic in todays world, a lot of problems have emerged. Of them the problem of traffic policemen to keep a check on riders for helmet has also become overwhelmingly difficult. We could see more vehicles on road than there are people on road. Everyone is in haste to reach his or her destination. So, policemen cannot stop every vehicle and check them for following traffic rules. This solution will be solving this problem of policemen and making it automatic.

There have been solutions like this in the developed nations but they are mainly focused on cars. Scenario in less developed countries is a little bit different. For instance, in India there are more two-wheelers than cars. And the major traffic rule for bike riders is with the helmet. Traffic rule instructs riders to wear helmet while riding the motorcycle but people think that wearing helmet is a tedious and uneasy task. People find it suffocating wearing the helmet during riding the motorcycle. Police check the vehicles mainly on intersections so riders take advantage of this. They wear the helmet on their hands, maybe to protect their elbows, and by the time they reach near the policemen they wear the helmet. It might be more tedious to carry the helmet in hands while driving, but riders find it easier to have it in hand rather than on head. People do not ponder over the point of wearing the helmet, they just overlook this rule. They wear the helmet just for the sake of showing it to the policemen. The riders think policemen the judges of a fashion show in which wearing helmet is mandatory.

This paper talks about the problems caused by not wearing helmet, the statistics behind it, and the lastly the implementation that this paper is mainly written for. This solution could prove to be a technology helping in leveraging the Intelligent Traffic Systems of Smart Cities.

2 The Concept of a Smart City

A city can be defined as an entity in which a large collection of individuals live and thrive in a social, material and cultural environment. A city encircles a group of amenities and resources which are housing, transportation, sanitation, utilities and communication [1]. The proper coordination and interaction between these amenities provided by a city, makes up what is called a city. With the advent of industrial revolution, the migration of rural dwellers to an urban setting has increased manifold. And this caused the population in a fixed area of a city to exponentially increase. Due to this a lot of forest areas surrounding the city area got cut down to accommodate more city dwellers. This urbanization increased the rate of growth of industries, so there demands for labour, and thus more labourers came to city to get job, thereby contributing to overpopulation.

All these causes deteriorated the serene environment that once existed when the rural and urban population was in balance. But now, a lot of problems have emerged, that are continuously degrading the lifestyle and health of the city dwellers. To overcome these problems, a smart solution is required to fight it. And here comes the concept of a Smart City. A Smart City is an umbrella term describing a bouquet of digital technologies working and interacting in order to make a better city and eliminating the causes of its deterioration. These technologies mainly comprise Internet of Things (IOT), Artificial Intelligence, Machine Learning, Blockchain, Edge Computing, Networking and Communication. The surveillance cameras, smart dustbins, sensors on sanitary outlets and industrial chimneys, sensors in vehicles, cell phones and many others act as devices to collect data from around the city. This huge amount of real-time collected data carry a lot of useful insights that can be studied and analysed to create a better solution to fight a city-deteriorating factor [2].

2.1 Challenges

Smart Cities contribute to development and enhancement of various aspects of life. These include but not limited to Waste Management, Intelligent Traffic Management, Environmental Pollution Control, Smart Lighting, Smart Roads, Better Healthcare, Forest Fire Detection, Early Earthquake Detection, Potable Water Monitoring, Smart Electricity Management, Smart Security and Emergency Facilities, Intelligent Shopping and Payment at POS, Smart Agriculture, Smart Energy Management and Intrusion Detection Systems. To incorporate these many technology applications and running them in sync with each other is another challenge that is experienced in deploying a Smart City. With great technology, comes a great deal of challenges with it. Smart City is no such exception. The major challenges [3] with Smart City are:

1. *Infrastructural resources*: The whole architecture of a Smart City employs so many technologies that have to run in sync with each other and be able to process information captured in real-time with least possible latency. For the accomplishment of this a lot of compute resources are needed which is expensive and not in the budget of every nation. Therefore, only some nations are accepting and implementing this technology. Before implementing these technologies, one should first ponder over power supply, cost incurred in maintenance and complications involved in running a city-wide technology.

2. *Security*: Wherever a remote connection, communication or internet is involved, there always exists a danger of security issues. Because the systems are monitored remotely via a channel that is connected to internet, it is vulnerable and open to attacks by hackers breaching the security of the intelligent systems infrastructure. So, tackling the security issues stands a major issue in implementation and maintenance of Smart City deployment. It will be very risky if the whole thread to puppeteer the entire city comes under the control of a malicious mind.

3. *Privacy of city dwellers*: With surveillance cameras installed, the privacy of individuals is snatched without consent. The cameras monitors behaviour and activities of every individual and this may just kill his or her privacy. But wherever it is required to collect data, privacy concerns arise automatically. To fight this, many organizations are working on to develop technologies that will enable systems to collect data but without harming or killing peoples right to privacy. Such technologies are Federated Learning and the concept of Differential Privacy [4, 5]. By using these two concepts the privacy concerns getting in the way of Smart City implementations can be done away with.

4. *Smart citizens*: For a Smart City to exist and function properly, it must be able to coexist and be in sync with its dwellers. The citizens must be educated and informed enough to adapt to these technologies and play a key role in better development of the city of future. This can be done through mass crowd-sourcing, citizen empowerment and email campaigns.

5. *Updated and socially inclusive population*: If half of the population, like elderly citizens, cannot afford to use mobile phones or applications, then the concept of smart city wont be able to spread much and benefit the city as a whole. A Smart City planning must involve these subtleties and consider all the citizens, and make them informed about or think another way out to make the technology accessible to those who cant access it using traditional means like mobile phone.

2.2   Smart Cities in India: A Case Study

The concept of Smart City has been employed by India under the Smart Cities Mission by the Ministry of Housing and Urban Affairs, Government of India. Under this plan India aims to conduct this mission under three components, which are, retrofitting (city improvement), redevelopment (city renewal) and greenfield development (city extension) [6]. India is an under-construction and fastly evolving country. It ranks second in terms of population. The waste management, traffic management, pollution management and other like issues are unable to be handled by the governmental organizations responsible for it. The population is ever increasing and the land area is that much only, so a lot of problems are arising due to large scale urbanization.

Through this programme, government wants to develop the country as well as fight the problems that are making Indias growth stagnant. Government conducted proposal submissions for smart city for various cities by various individuals and organizations. The government is implementing the proposals in some major cities.

Fig. 1 Dholera SIR is the first smart city in India located in the state of Gujarat. There are more smart cities to come up in the future as per the government's plans [7]

The problem of citizen empowerment is highly lacking in India. So the challenge of benefiting each and every citizen adversely affects the spread of this concept around the nation.

The traffic management in India is not very good. People dont want to follow traffic rules. When traffic policemen stop them for fines, they shut their mouth with bribe. The traffic policemen is also corrupt so he also accepts the bribe and lets the rider go and make his life or someone others life a hell. Not wearing a helmet while driving is thought to be cool. Moreover, people give the excuse of excessive hot weather for not wearing helmet. But is it better to not wear helmet and tolerate some heat than die on road without helmet by head injury? Absolutely not at least in theory, but it needs to be implemented in practice. Every citizen should not be just responsible for him or her but also think in advance what his or her wrong deeds on road will do to others on road. But people dont understand till they get someone to monitor their deeds. So the implementation that this paper talks about helps to keep a check on the riders for helmet and penalize them for not wearing one. The system also contributes to a better Intelligent Traffic Management System in India (Figs. 1 and 2).

3 Motivation

One of the major concerns of Smart Cities is the traffic management. Smart Traffic Management includes solutions providing commuters with facilities like easily finding a parking space using a mobile phone. But Smart Traffic Management does

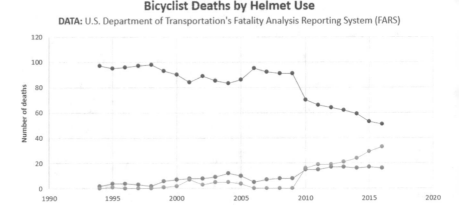

Fig. 2 Statistics showing the effect of wearing a helmet during an accident (*Data Source* [9])

not only concern with providing facilities to its citizens, but also to keep a check on its dwellers that they follow the traffic rules properly, so that no issue may occur in proper functioning of the Smart City. With smarter city and more urbanisation, the vehicles on roads will increase and checking them for following traffic rules manually would be too cumbersome for the traffic policemen. So, there is a need for a solution that could check riders and drivers for following traffic rules properly.

Driving a motorcycle on road is more vulnerable than driving a car or other heavy vehicle [8]. This is due to the fact that in the case of any collision or accident, the motorcycle riders body is more open to impact than other vehicles like cars in which the driver is inside a case, which can protect the driver to some extent. According to the World Health Organization (WHO), wearing helmet reduces head injury by around 20–45%, and those not wearing a helmet while riding are three times more prone to head injury than those wearing a helmet. Young passengers rarely wear a helmet and on the top of it, they want to flaunt and over-speed. It is also evident that in countries where helmet wearing laws are strict, the injuries caused by accident to two-wheeler riders has decreased by 20–30%. The implementation of strict helmet wearing laws in Italy increased the helmet wearing from 20% in 1999 to more than 96% in 2001, and decreased the deaths and head injuries by a huge margin. Laws implemented in Malaysia reduced the head injuries by 30% [10]. As per the National Highway Traffic Safety Administration studies, riders who were not wearing helmet had higher health care costs, longer hospitalization, longer recovery periods, and more severe disorders or disabilities [11].

Mostly people do not wear helmet because they feel hot and suffocating to wear the helmet. But, feeling hot for a small duration is better than death. Moreover, riders should buy good quality helmets with proper air vents and tie them properly for optimum security. There are many excuses people give for not wearing a helmet, but according to Nicole Levy, MD, a primary care sports medicine specialist at Rush

University Medical Center, there exist no valid excuses for not wearing a helmet. A serious injury can occur to the brain as a result of the accident. The helmet absorbs most of the impact during the accident and saves the rider from serious consequences of brain trauma. A serious brain injury can lead to skull fracture, lost consciousness, memory loss, concentration loss, sleep disorders and even death [12].

4 Concept

The solution that this paper talks about will be able to detect riders for wearing helmet and penalize them if they are found not wearing helmet. The solution is built using YOLO and darknet framework and was trained with *TensorFlow* as backend. The model is trained on custom manually annotated image dataset and trained on two-GPU server for 2500 epochs.

The detection of faces and number plates were performed by traditional machine learning methods for computer vision available in OpenCV like Haar Cascade for face detection and, Canny Edge Detection and Contour Detection for number plate detection.

The implementation works in two phases. In the first phase, the riders are detected for helmets. If the rider is found to be wearing a helmet and his face is detected or not, then he is let go. If the rider is found to be not wearing the helmet, and his face is detected, that means he is without helmet then another phase of the solution will start. In the second phase, the number plate of the rider who is found to be not wearing a helmet, is captured and read. Then the riders motorcycle number is stored in a server. Afterwards, a fine notice is generated automatically and is sent to the address of the rider [13] (Fig. 3).

The implementation employs three major modules. One for helmet detection, other for face detection and the last for Automatic Number Plate Recognition (ANPR). The face detection is performed using OpenCV and Helmet Detection is performed using trained YOLO model and OpenCV. The Automatic Number Plate Detection (ANPR) is performed using the API from Plate Recognizer [14]. And these detections go on in real time. It is like binary classification at first step and then conditional steps afterwards. The flowchart representing the implementation flow plan is as depicted in figure above.

4.1 Object Detection

It is a task in computer vision and image processing that aims to detect objects in a digital image, video or real-time camera feed. It detects objects for which it is trained for. It has many applications in areas of computer vision like image retrieval, pedestrian detection, video surveillance, and face detection. Nowadays it is being used in many other tasks like fault detection through image, behaviour detection and

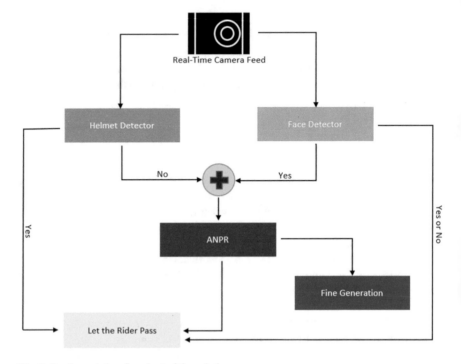

Fig. 3 Implementation flowchart of the solution

object segmentation. This solution is also an application of object detection that will be using real-time traffic camera feeds.

There exist two approaches for performing object detection, which are machine learning approach and deep learning approach. In machine learning approach, features have to be explicitly specified and defined. These use one algorithm for extracting features from the image and another for classification of the detected object based on the features. For feature extraction, the algorithms commonly used are *Viola-Jones* algorithm based on *Haar features*, Scale invariant feature transform (*SIFT*) and Histogram of oriented gradients (*HOG*). For classification mostly Support Vector machines (*SVM*) is used.

4.2 YOLO

YOLO or *You Only Look Once* is an object detection framework based on darknet. Darknet is a framework written in C that performs object detection on images, videos and camera feeds. It is now in its third iteration (YOLOv3). Though many object detection frameworks exist in the market, YOLO is one of the most widely used and the reliable one. It is always updated with its codebase and pre-trained weights

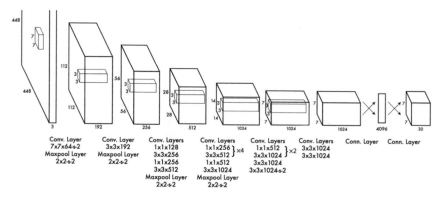

Fig. 4 YOLO architecture [15]

that one can use if one wants to train a custom object detector or use pre-trained weights to make a general object detector. The default weights and configuration files provided by the YOLO are trained to detect almost all the commonly used objects and these objects are included by Microsofts Common Objects in Context (COCO) objects dataset. The YOLO models come in various shapes and sizes; they have different number of layers that corresponds to better mean average precision(mAP). This model is most widely used nowadays because of its performance along with least inference time per frame or image. YOLO gives better mAP as well as has least inference time in comparison to its closest competitor Facebook AI Researchs RetinaNet (Fig. 4).

In deep learning approach, it is not required to explicitly define and specify the features. These employ convolutional neural networks (CNN) that are competent enough to create and learn features. The commonly used techniques to perform the task of object detection are *R-CNN*, *Fast R-CNN*, *Faster R-CNN*, Single Shot Multi-box detector (*SSD*), You Only Look Once (*YOLO*), Facebook AI Researchs *RetinaNet* and *TensorFlow Object Detection API*.

YOLO works by passing the frame of the video or image through a single neural network that divides the image into different regions and calculates bounding boxes with some probabilities. The bounding boxes with most probability for the desired class is chosen [15].

Other detectors like RCNN and Fast RCNN calculate bounding boxes for different portions of the image on different scales. So, these algorithms calculate many, around thousands of bounding boxes for detecting a single object. It is due to this fact that YOLO performs around 1000 times faster than RCNN and 100 times faster than Fast RCNN.

4.3 Curse of Dimensionality

This is an inherent problem with machine learning and neural networks. There exists a trade-off between number of features in the data and number of epochs for which the model is trained on the data. Although this term has deep roots emanating from Pure Mathematics, yet it has different interpretations, one of which is used in reference to machine learning algorithms. Bellman coined this term to describe the problem of exponential increase in volume when a lower dimensional Euclidean space is taken to higher dimensions [16].

In reference to machine learning, dimensionality refers to the number of features that are there in the data. It comes in to play during the training process of the algorithm. When the training starts, at this point of time, the model has learnt very little and is not yet capable enough to perform its desired task, that is it is a very simple approximation of the original to-be model. This is known as underfitting. After some number of epochs or iterations over the entire dataset, the models best state is achieved after which the models accuracy stalls, precision and recall start approaching 100%. This state of the model is known as sweet spot and this is the state which every model aims for. After this spot, any increase in epochs dwindles the accuracy and generalization power of the model in real-world scenarios and unseen cases. This unwanted state is known as overfitting of the model [17] (Fig. 5).

In the case of underfitting, the model is too simple and too biased, while in the case of overfitting, the model produced is too complex and its variance is too high. In other words, it can be said that an underfitting model learns little to nothing about the data it is trained upon, and an overfitting model tends to learn too much about the data, that is it just memorizes the data, so it becomes useless to be used on unseen data. So, in both cases validation error is too high, though training error may be small in the case of overfitted model [18].

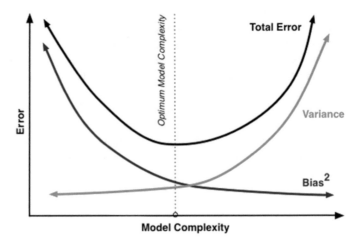

Fig. 5 Bias-variance tradeoff

Fig. 6 Annotating image using the open-source LabelImg tool

But this problem of dimensionality can be dealt with by using regularisation. Regularisation is a technique that helps to downgrade the complexity of a model, so that the element of overfitting from the model gets removed. By this the model approaches the sweet spot and a better model is achieved. Regularisation is an umbrella term for various techniques. These techniques are ln-regularisation, data augmentation, early stopping, batch normalization, and dropout. These techniques all work to reduce to find the sweet spot of the model. Of them, dropout has been considered the best for most of the cases till now. While data augmentation is a technique that is mostly used in the case of computer vision to augment the data by modifying the images, that is by rotation, colour changing, reverting, etc. Early stopping as the name suggests stops the algorithm from further training when that sweet spot is found (Fig. 6).

5   Preparing the Data and Training

The data for the implementation was created manually. First the images from Google Image search for keyword 'helmet' and 'traffic', and 100 images related to these keywords were downloaded. The model had to be trained on a one-class data, that is "Helmet". Thereafter, the images were annotated and marked to specify the desired portions of the images for detecting helmet. The data was labelled using *LabelIMG* tool which is a free open source tool to label and annotate images. This tool provides feature of making bounding boxes on desired portions of image and can generate annotations of two types YOLO and Pascal VOC format [19].

After the annotation step, the whole directory of *darknet* was configured to meet the needs of the required model. The weights chosen for training the model was *darknet53.conv.74*. The configuration file was identical to yolov3 configuration file. [YOLO] The model was trained for over 2500 epochs on dual *Nvidia Quadro P6000*

GPU server. It took around 25 min for training to complete. The training procedure was halted after 2500th epoch because on reaching around 2500 epochs, the model seemed to start overfitting and since a real time general model is required so overfitting is a curse for this application.

6 Results and Discussions

The system is able to perform detection of helmets in static images, videos and real time camera feeds. The detector is made using OpenCV to implement all the required utilities for detecting of helmets [20]. I chose OpenCV over the usual, original darknet detect utility, because of the speed and latency factor. OpenCV is faster than darknets detect because it performs much less calculations and moreover since it uses CPU it is less hardware-resource demanding. And these all benefits make OpenCV a good choice for real-time detection.

The output image or image or camera feed from the helmet detector module shows detected bounding boxes (in orange) with the label Helmet along with confidence percentage with which the trained model detected the object, that is helmet.

The automatic number plate recognizer(ANPR) module uses the API from Plate Recognizer. The detector module sends a request of the image or video frame being fed into it, and receives a json response. The field containing vehicle plate number in the response is then extracted using normal Python methods.

The face detector module utilizes the *Haar Cascade features* available in OpenCV [20]. The image is first converted to grayscale and then face is detected using Multi-scale face-cascade algorithm.

The detection procedure on an image or a pre-recorded video is a little slow, and of course is not performed in real time, but this can be extended to real time using OpenCVs VideoCapture() function. Using this the system can be deployed along with a real time recording camera, to make the camera intelligent enough to detect helmets in the feeds it captures.

The helmet detection module and the face detection procedure, both work simultaneously, and return boolean results for detections for a frame of a camera feed. After receiving results from these two procedures, the further processing is decided. If helmet is detected, then irrespective of the result of face detection, the rider is let to pass without any fine. If the helmet is not detected and a face is detected, then that means the rider is present in the video frame, but rider is not wearing a helmet. In this case, number plate in the current video frame is detected and stored in the traffic fine database. This stored fine database information is then used to generate fine automatically based on the vehicle registration information (Fig. 7, 8 and 9).

The results are demonstrated using a Flask web application [21]. The application is a two-page application, one to upload and save the image or video or camera feed, and the other to show the results. The experiment was performed on three types of images

Fig. 7 Result of helmet detector module on an image

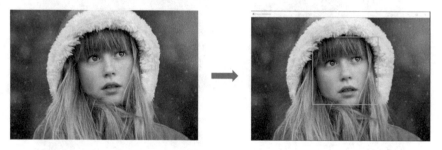

Fig. 8 Result of face detector module on an image

1. With rider having helmet but his or her face could not be seen.
2. With rider having helmet with his or her face being seen.
3. With rider not wearing a helmet, and his or face being seen.

As can be seen in Figs. 10, 11 and 12, the system performs well on the three use cases that have been considered in making this application. This system has a lot of potential to better contribute for developing an Intelligent Traffic Management System in a Smart City.

This application or its like can be deployed on a secure environment and connected to Traffic Police database wherein different analysis can be carried out on the data collected [22]. The camera feeds from the traffic surveillance cameras will be fed into the application in real-time, and then the further procedures will be carried out - like letting the rider go or automatically penalizing the rider for not wearing a helmet by generating a fine for the rider's motorcycle registration number. The cameras can use IOT-enabled smart sensors which could just manipulate the recorded camera feeds in real time thus decreasing latency and increasing the overall response time of the system.

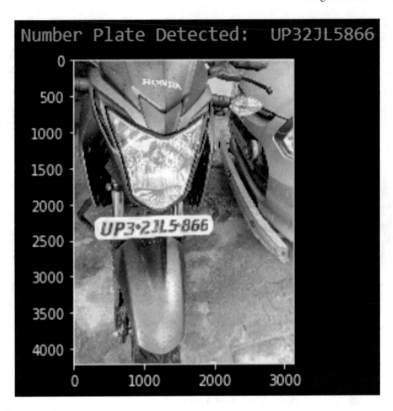

Fig. 9 Result of ANPR module on an image

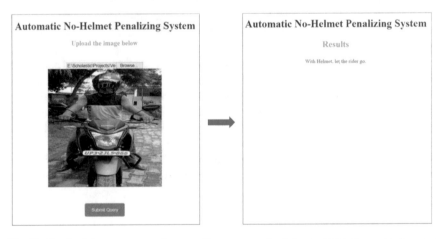

Fig. 10 Result of the system on an image with person wearing helmet with his face not being seen

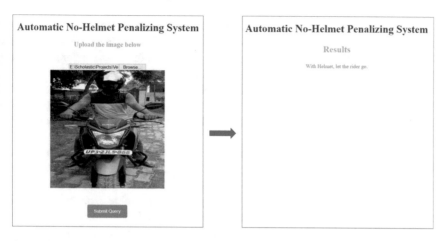

Fig. 11 Result of the system on an image with person wearing helmet with his face being seen

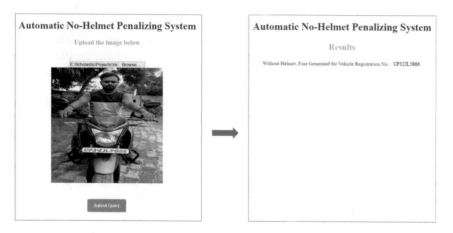

Fig. 12 Result of the system on an image with person without helmet

7 Future Work

This new century has embarked on the path towards the Fourth Industrial Revolution, wherein technologies like Artificial Intelligence (AI), Internet of Things (IOT) and Blockchain will be helping in building a smart city - a city of the future. A city where each and every entity whether living, non-living or abstract will be interacting with each other to enable a smart planet in totality. Smart and intelligent systems will be able to sense and gain useful insights from the data collected from IOT sensors. These systems will be utilising the power of AI to instil the smartness of a human being to an abstract of non-living entity. These systems could be made more secure using the Blockchain paradigm.

The system we talk about here could serve to leverage the services of an Intelligent traffic management system deployed in a smart city [23, 24]. Using this system, we will be able to enforce the traffic rules regarding helmets in a much better automatic way. Here we used the AI paradigm and IOT paradigm. The sensors including the cameras which will convey the traffic surveillance data to a server leverages IOT [25], and the analysis that will take on the server is the task of AI. But we did not talk about the third paradigm that we mentioned above - the possible uses of Blockchain in this application.

Blockchain is one of the future technologies. The notion of it came into existence when Satoshi Nakamoto presented a paper in 2008 that aimed to solve the problem of double spending in transactions without the need of a mediator or a financial institution like bank [26]. In simple terms they wanted to enable exchanging money between two entities in a secure manner without each needing a bank between them.

That was the story regarding financial transaction. In computer science, transaction is defined in layman-terms as a collection of tasks that must be completed in order for a process to accomplish. Since a blockchain is concerned with providing a decentralised base to a transaction and securing it on top of it, it can be used in our application as well. Our system can use blockchain to secure the communications between the smart surveillance cameras and the processing server. The transactions will be the data packets that would be sent by the camera systems to the database for processing [27].

Moreover, these servers can be done away with by employing a full blockchain network of smart surveillance cameras. The cameras could then be embedded with smart sensors and edge computes. These cameras will be smart enough to process the camera feeds themselves, and this information can be then verified by the blockchain of cameras, where each camera will be a block, and transactions will be the decision of penalising a rider. Through this, the concept of an Intelligent Traffic Management System would be fully successful in terms of technology stack utilisation and implementation.

Since we are talking about applying blockchain to our system, we can not leave the concept of incentives here. The riders who are wearing helmets can be given rewards by the blockchain network [28, 29]. Through this, the message of wearing helmets will spread nation-wide. Every rider will then have no chance to bribe a traffic policemen or hack into the blockchain system, thereby providing a safe, secure and intelligent surveillance system.

Acknowledgements This research was supported by Hankuk University of Foreign Studies Research Fund and VESTELLA Labs Inc. We thank our colleagues for providing valuable insights and expertise that greatly assisted in the completion of this work.

References

1. McFedries P (2014) The city as system (Technically Speaking). IEEE Spectrum 51(4):36. https://doi.org/10.1109/MSPEC.2014.6776302
2. Suzuki LR (2017) Smart cities IoT: enablers and technology road map. In: Rassia S, Pardalos P (eds) Smart city networks. Springer Optimization and Its Applications, vol 125. Springer, Cham. https://doi.org/10.1007/978-3-319-61313-0
3. Sydney Stone, Key challenges of Smart Cities and how to overcome them. https://ubidots.com/blog/the-key-challenges-for-smart-cities/
4. Dwork C (2006) Differential privacy. In: ICALP, pp 1–12. https://doi.org/10.1007/11787006_1
5. McMahan B, Ramage D, Federated learning: collaborative machine learning without centralized training data. https://ai.googleblog.com/2017/04/federated-learning-collaborative.html
6. Smart Cities Mission, Government of India. http://smartcities.gov.in/content/
7. Dholera SIRDA. http://dholerasir.com/
8. Bourdet N, Deck C, Tinard V, Willinger R (2012). Behaviour of helmets during head impact in real accident cases of motorcyclists. Int J Crashworthiness 17:51–61. https://doi.org/10.1080/13588265.2011.625676
9. Bicycle Helmet Safety Institute. https://helmets.org/stats.htm
10. Helmets: a road safety manual for decision-makers and practitioners. https://www.who.int/roadsafety/projects/manuals/helmet_manual/en/
11. Why is it so important to wear a helmet on motorcycles? https://mainorwirth.com/blog/why-is-it-so-important-to-wear-a-helmet-on-motorcycles/
12. Helmet Safety: Keep a Lid on it. https://www.rush.edu/health-wellness/discover-health/helmet-safety-keep-lid-it
13. Rehman GU, Ghani A, Zubair M, Naqvi SHA, Muhammad S, Singh D (2019) IPS: incentive and punishment scheme for omitting selfishness in the internet of vehicles (IoV), IEEE Access, August 2019. https://doi.org/10.1109/ACCESS.2019.2933873
14. Plate Recognizer API. https://platerecognizer.com/
15. Joseph Redmon, Ali Farhadi; YOLOv3: An Incremental Improvement. https://arxiv.org/pdf/1506.02640
16. Sammut C, Geoffrey I. Webb (eds) Encyclopedia of machine learning, Springer Reference. https://doi.org/10.1007/978-0-387-30164-8
17. Goodfellow I, Bengio Y, Courville A, Deep learning. MIT Press. http://www.deeplearningbook.org/
18. Burkov A, The hundred-page machine learning book. http://themlbook.com/
19. Darrenl, LabelImg image annotation tool. https://github.com/tzutalin/labelImg
20. OpenCV: an open-source computer vision and machine learning library. https://opencv.org
21. Flask: a web application framework in Python. https://palletsprojects.com/p/flask/
22. Yadav P, Jung S, Singh D (2019) Machine learning-based real-time vehicle data analysis for safe driving modeling. In: The 34th ACM/SIGAPP symposium on applied computing (SAC), Limassol, Cyprus, April 8–12, 2019, pp 1355–1358. https://doi.org/10.1145/3297280.3297584
23. Singh D, Singh M (2015) Internet of vehicles for smart and safe driving. In: International conference on connected vehicles and expo (ICCVE), Shenzhen, 19–23 Oct., 2015. https://doi.org/10.1109/ICCVE.2015.93
24. Singh D (2013) Developing an architecture: scalability, mobility, control, and isolation on future internet services. In: Second international conference on advances in computing, communications and informatics (ICACCI-2013), Mysore, India, August 22–25, 2013. https://doi.org/10.1109/ICACCI.2013.6637467
25. Singh D, Tripathi G, Jara A (2015) Secure layers based architecture for internet of things services. In: IEEE world forum on internet of things (WF-IoT), Milan, Italy, Dec. 14–16, 2015. https://doi.org/10.1109/WF-IoT.2015.7389074
26. Nakamoto S (24 May 2009) Bitcoin: a peer-to-peer electronic cash system. https://bitcoin.org/bitcoin.pdf

27. Singh M, Kim S (2018) Branch based blockchain technology in intelligent vehicle. Comput Networks ISSN -1389-1286. https://doi.org/10.1016/j.comnet.2018.08.016 (IF 2.522)
28. Singh M, Kim S (2017) Introduce reward-based intelligent vehicle communication using blockchain technology. In: 14th international SoC design conference (ISOCC 2017), Grand Hilton Hotel, Seoul, South Korea, Nov. 5–8, 2017. https://doi.org/10.1109/ISOCC.2017.8368806
29. Singh M, Kim S (2018) Trust bit: reward-based intelligent vehicles communication using blockchain. In: The 4th IEEE world forum on the intelligent of things (WF-IoT), Singapore, Feb. 05–08, 2018. https://doi.org/10.1109/WF-IoT.2018.8355227

An Overview of Smart City: Observation, Technologies, Challenges and Blockchain Applications

Vijay Kumar Chaurasia, Alhasha Yunus and Madhusudan Singh

Abstract Smart cities are becoming smarter because of the currently development of technology world. A smart city includes different electronic devices such as street cameras, sensors for transportation system, GSM module for smart waste management, etc. Blockchain Technology in smart cities is provide efficient secure peer to peer network in huge data world in those generated in smart cities use cases such as healthcare data, autonomous vehicles communication environment. Smart city technologies are encouraging the use of smart phones to connect with everything's, and also person can access the things data through the smart phone. Therefore, Internet of Things (IoT) are playing a great role in making the cities smarter. This chapter aims to provide a comprehensive overview of the use of smart city technologies and its application challenges. This chapter will further conclude about the smart systems which is going to install in smart cities to reduce the human efforts and ensure more security and ease to human beings.

Keywords Smart cities · IoT · Blockchain · Digital technologies · Connected vehicle

1 Introduction

A smart city is a city or urban area that makes use of different IoT applications. A smart city uses sensors to collect data and send it to control center to manage the actions efficiently. This includes data, which is collected from citizens, devices,

V. K. Chaurasia
Department of IT, Indian Institute of Information Technology, IIIT-A, Allahabad, India

A. Yunus
Research Intern, IIIT-A Campus, MtoV Inc, Seoul, South Korea

M. Singh (✉)
School of Technology Studies, Endicott of International Studies, Woosong University, Daejeon, South Korea
e-mail: msingh@wsu.ac.kr

© Springer Nature Singapore Pte Ltd. 2020
D. Singh and N. S. Rajput (eds.), *Blockchain Technology for Smart Cities*,
Blockchain Technologies, https://doi.org/10.1007/978-981-15-2205-5_7

cars, streetlights, plants, homes etc. This data is processed and analyzed to organize the smart transportation system, home automation, smart pollution control system, weather forecasting, smart healthcare system, smart traffic management system, smart gardens etc. It uses internet as a medium to communicate with the objects or devices. 5G technologies are used in smart cities for better flow of communication between sensors and devices. It helps in giving real time data and responses so that fault occurrence can be reduced [1]. The 5G technology is the next generation of wireless communications. It is expected to provide Internet connections that are least 40 times faster than 4G LTE. 5G technology may use a variety of electromagnetic bands including millimeter wave, which can carry very large amounts of data at short distance. The drawback of the higher frequencies is that they are more easily obstructed by walls, trees and other foliage, and even inclement weather [2].

In modern cities, people do not have time to look after their gardens therefore automated gardening can be used to provide better facilities at homes. Plants are very beneficial to the human beings in many ways for example it can refresh one's mind and soul. Plants help in keeping the environment clean by cleaning air naturally and producing oxygen. Excessive watering and less watering, both are harmful for plants. Plants require a specific amount of sunlight, water and temperature to remain healthy and growing. Using automated system, we can control these factors and plants can grow in a better way [3]. Streetlight automation in smart cities is a very efficient approach for reducing the consumption of energy and provides more security to pedestrians and vehicles [4]. In cities, we are facing accidents on roads very often. To prevent the accidents, it is necessary to implement an intelligent transportation system that can intelligently interact with vehicles. This system is called as Intelligent Transportation System (ITS) [5]. Blockchain technology can help to secure protect, verified secure data in Smart cities digitization. Our environment is getting polluted day by day because no proper solid waste management system is in place which is creating a bad impact on the health and environment of our society. Smart city technologies have a solution for this which is known as garbage automation. Garbage automation helps in detecting, monitoring and managing wastes automatically. And helps in keeping the cities clean and healthy [6]. The smart city comprises of all these technologies and helps in providing more safety to people living there. Internet of things has a major role in implementing these ideas. Due to advancement in technologies, smart cities are getting smarter. IoT based applications are extremely necessary for the expansion of smart cities. As everything is connected to internet and every IoT device is generally controlled by mobile phones, it's simple for user to access information from any place. Smart cities embody smart energy, smart inhabitants, smart care, smart transit, smart security etc. The revolution of internet is acting as the backbone for smart cities. Smart city technologies are influencing the cities in many diverse ways. Smart city technologies provide more organized, economic and secure operation of the system constructed on various aspects like energy saving policies, economic cogitation, reliability levels, etc. [7].

In the IoT context, devices are often integrated based on the geographic location and evaluated by analyzing system. Sensors for gathering specific data can be used in smart city applications. There are several application domains that uses

IoT infrastructure to facilitate operations like air and sound pollution, the quality of vehicles and the inspection systems. The revolution of the internet provides an infrastructure during which many people are able to interconnect to every alternative. Subsequent revolution of the communication technology can facilitate appropriate interconnections among the objects. In the coming years, the number of objects that are interconnected will be far more than the number of individuals [8]. On one hand, IoT can have an impact on the assorted aspects of the citizens' life in the field of health, security, and transportation on the other hand, it can play a very important role at the national level in making the policy decisions like energy saving, pollution decrement, remote observance and infrastructure monitoring. On this basis, IoT will facilitate the supply a lot of systematic, economic and secure operation of the system based on totally different aspects, like energy saving policies, economic issues, reliability levels, etc. Smart cities use information and technology to form efficiencies, improve viability, produce economic development, and enhance quality of life factors for individuals living and working in the city [7]. Smart city uses information technology to interact effectively with native people in native governance and decision by use of open innovation processes and e-participation, raising the collective intelligence of the city's establishments through e-governance, with emphasis placed on subject participation and co-design. Smart city employs a mixture of data assortment, processing, and dispersive technologies in conjunction with networking and computing technologies and data security and privacy measures encouraging application innovation to push the quality of life for its citizens and covering dimensions that include: utilities, health, transportation, recreation and government services.

2 Smart City Technologies

Smart city technologies are playing a vital role in the growth and success of smart cities. Some of the technologies applied in smart cities are discussed in the following (Fig. 1).

2.1 Wireless Sensor Network (WSN)

Wireless sensor network (WSN) refers to a class of spatially spread and dedicated sensors for observing and recording the physical parameters of the surrounding by organizing the collected information at a central location. WSNs evaluate environmental conditions like temperature, sound, pollution levels, humidity, wind, and so on. These are just wireless ad hoc networks that depend upon wireless connectivity and spontaneous network formation so that sensor data will be transported wirelessly. WSNs are using spatially distributed autonomous sensors to observe physical or environmental conditions. Modern networks are bi-directional, additionally sanctioning management of sensor activity. The development of wireless sensor networks was

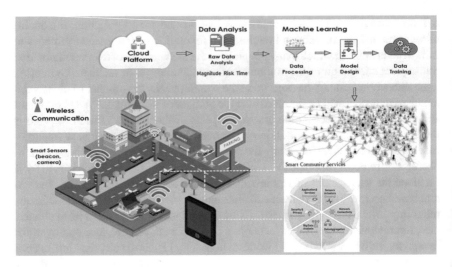

Fig. 1 Overview of smart city system

impelled by military applications like tract surveillance; nowadays such networks are utilized in several industrial and client applications, like process observation and management, machine health observation, and so on. The WSN is a network of sensor nodes whose number may vary from few nodes to many hundreds or maybe thousands, where each node is connected to 1 or sometimes several sensors. Each such sensor network node has generally many components, a radio transceiver with an enclosed antenna or association to an external antenna, a microcontroller, and an electronic circuit for interfacing with the sensors and an energy supply, typically battery or an embedded kind of energy harvesting. A sensor node would possibly vary in size from that of a shoebox down to the scale of a grain of dust, although functioning modes of real microscopic dimensions have yet to be created. The value of detector nodes is similarly variable, starting from a few to many dollars, betting on the complexness of the individual sensor nodes. Size and price constraints on sensor nodes lead to corresponding constraints on resources like energy, memory, computational speed and communications bandwidth. The topology of the WSNs can vary from an easy star network to a complicated multi-hop wireless mesh network. The propagation technique between the hops of the network may be routing or flooding.

In area observance, the WSN is deployed over an area where some phenomenon is to be monitored. A military example is the use of sensors to observe enemy intrusion; a civilian example is the geo-fencing of gas or oil pipelines. There are many varieties of detector networks for medical applications: implanted, wearable, and environment embedded. Implantable medical devices are those that are inserted within the physical body. Wearable devices are used on the body surface of an individual or simply at close proximity of the user. Environment-embedded systems use sensors contained within the environment. Possible applications embody the body position measurement, location of persons, and overall observation of sick patients in hospitals and at home. Devices embedded in the environment track the physical state of an individual for continuous health diagnosis, using as input the information from a network of depth cameras, a sensing floor, or alternative similar devices. Body-area networks will collect information concerning a person's health, fitness, and energy expenditure. In health care applications, the privacy and legitimacy of user data has prime importance especially because of the mixing of sensor networks, with IoT, the user authentication becomes more challenging. Wireless sensor networks are deployed in several cities like Stockholm, London, and Brisbane to observe the concentration of dangerous gases for citizens. These will take advantage of the unplanned wireless links instead of wired installations, which additionally make them mobile for testing readings in several areas. Operative systems for wireless sensor network nodes are generally less advanced than general-purpose operating systems [9]. They more powerfully resemble embedded systems, for 2 reasons. First, wireless sensor networks are usually deployed with a specific application in mind, instead of as a general platform. Second, a requirement for low prices and low power leads most wireless sensor nodes to have low-power microcontrollers ensuring that mechanisms like virtual memory are either unnecessary or too costly to implement.

2.2 Radio-Frequency Identification

Radio-frequency identification (RFID) uses electromagnetic fields to mechanically determine and track tags connected to things. The tags contain electronically saved information. RFID tags can be passive, active or battery-assisted passive. The active tag has an on-board battery and sporadically transmits its ID signal. Active tags have a local power supply and will operate hundreds of meters from the RFID reader. A battery-assisted passive contains a little battery on board and is activated in the presence of an RFID reader. A passive tag is cheaper and smaller as a result of it's no battery; instead, it uses radio energy transmitted by the reader. However, to control a passive tag, it should be lighted with a power level roughly a thousand times stronger than for signal transmission that produces a distinction in interference and in exposure to radiation. In contrast to a barcode, the tag needn't to be inside the line of sight of the reader, thus it may be embedded in the tracked object. The tag is read if passed close to a reader, even if coated lines of object are not visible. The tag is scanned from within a case, carton, box or alternative container, and in contrast to barcodes, RFID tags are read a whole lot at a time. Bar codes will solely be scanned one at a time using current devices. RFID is the technique of automatic identification and data capture. Tags could either be read-only, having a factory-assigned serial number that's used as a key into a database, or is also read/write, wherever object-specific data is written into the tag by the system user. Field programmable tags can be written once and read multiple times, blank tags can be written with an electronic product code by the user. RFID tags contain a minimum of 3 parts, an integrated circuit that stores and processes data which modulates and demodulates radio-frequency signals; a way of grouping DC power from the incident reader signal; and an antenna for receiving and sending the signal. The tag information is saved in a non-volatile memory. The RFID tag includes either fixed or programmable logic for processing and transmission of sensor information respectively [10].

An RFID reader transmits an encoded radio wave to interrogate the tag. On receiving the message, the tag responds with its identification and alternative information. This might be solely a unique tag serial number or may be product-related data like a stock number, lot or batch number, production date, or other specific information. Since tags have individual serial numbers, the RFID system can discriminate among them which might be inside the range of RFID reader and skim them at the same time. RFID systems are classified by the sort of tag and reader. A Passive Reader Active Tag system contains a passive reader that only receives radio signals from active tags. The reception range of a prat system reader can be adjusted from 1 to 2,000 ft, permitting flexibility in applications like asset protection and management. An Active Reader Passive Tag system has an active reader, which transmits interrogator signals and conjointly receives authentication replies from passive tags. An Active Reader Active Tag system uses active tags awoken with an interrogator signal from the active reader. A variation of this system might conjointly use a Battery-Assisted Passive tag that acts as a passive tag but contains a little battery to power the tag's return reporting signal. Fixed readers are set up to form a particular interrogation zone which can be tightly controlled. This permits an extremely defined reading space for tags when they enter and exit from the interrogation zone. Mobile readers are also hand-held or mounted on carts or vehicles. Signaling between the reader and the tag is completed in several different incompatible methods, based on the frequency band utilized by the tag. Tags operating on LF and HF bands are radio wavelengths which are very close to the reader antenna because they're solely a little percentage of a wavelength away. In this close field region, the tag is closely coupled electrically with the transmitter in the reader. The tag will modulate the field created by the reader by changing the electrical loading which it represents by changing between lower and higher relative loads, the tag produces a modification that the reader will observe. At UHF and higher frequencies, the tag is more than one radio wavelength far away from the reader, requiring a distinct approach. The tag will backscatter a signal. Active tags might contain functionally separated transmitters and receivers, and they needn't to respond to any frequency associated with the reader's interrogation signal. The RFID tag is pasted to an object to track and manage inventory, assets, people, etc. for example it can be pasted on cars, laptop equipment, books, mobile phones, etc.

2.3   5G Technologies

5G networks are digital cellular networks within which the service area coated by suppliers is split into little geographical areas referred to as cells. Analog signals representing sounds and pictures are digitized in the phone changed by an analog to digital converter and transmitted as a stream of bits. All the 5G wireless devices in a cell communicate by radio waves with an area antenna array and low power automatic transceiver within the cell over frequency channels allotted by the transceiver from a standard pool of frequencies that are reused in geographically separated cells. The native antennas are connected with the telephone network and the internet by a high bandwidth fiber or wireless back haul connection. Like existing cell phones, when a user crosses from one cell to a different, their mobile device is automatically handed off seamlessly to the antenna in the new cell. In the Internet of Things, 3GPP is going to submit evolution of NB-IoT and eMTC (LTE-M) because of the 5G technology for the Low Power Wide Area use case [11].

2.4 Cloud Computing

Cloud computing is the accessibility of computer system components as and when it required. It can be data storage and/or computing power. It is not necessary for the user to actively manage these things on cloud. The term cloud computing is generally used to refer data centers that are available to the users on the Internet. The central servers assign different functions to the multiple locations controlled by large clouds. If the connection to the person who is using it is quite close, it can be labeled as an edge server. Cloud computing depends fully on sharing of resources to accomplish logicality, consistency and proportionate saving in costs. Cloud computing permits companies to evade or reduce up-front IT infrastructure costs. It is claimed by proponents that cloud computing permits organizations to get their applications up and run more rapidly with enhanced manageability and reduced maintenance and allows IT teams to adjust resources faster to meet unstable or uncertain demand [12].

The main aim of cloud computing is to enable users to gain profit from all of these technologies, even if the user has no knowledge about these technologies or he is not expert in these. The cloud aims to lower the costs and makes it easier for the users to accentuate their core business without being hindered by IT barriers. Virtualization is the basic technology which enables the cloud computing. Virtualization software splits physical computing device into different virtual devices, which can be used and managed comfortably to carry out computing functions. Virtualization provides the ability to work quickly and easily and helps in speeding up IT operations and lowers the cost by ascending infrastructure utilization. Autonomous computing automates the process through which user can equip resources whenever required. By reducing user participation, automation increases the speed of the process While reduce the labor cost and the probability of human error. Cloud computing utilizes ideas from utility computing to provide standard of measurement for the services utilized. Cloud computing aims to address the quality of service and reliability issues of grid computing models.

2.5   *Big Data*

Big data comprises of data sets with sizes which is difficult to capture, curate, manage, and process by commonly used software tools within a tolerable elapsed time. Big data philosophy encloses unstructured, semi-structured and structured data but the main attention is given to the unstructured data. Big data size is a continuously moving target. It ranges from a few dozen terabytes to many zettabytes of data. Big data need a set of techniques and technologies with latest forms of integration to divulge insights from datasets that are distinct, complex, and of a larger scale. Big data depicts the information assets distinguished by a high volume, velocity and variety to require particular technology and analytical methods for its modification into value. When parallel computing tools are needed to handle data then big data comes into action. The challenges encompass capture, storage, search, sharing, transfer, analysis, and visualization [13].

Big data and the Internet of things work in conjunction. Data extricated from IoT devices supply a mapping of device interconnectivity. These mappings are utilized by the media, enterprises and governments to target their audience more appropriately and increase media efficiency. IoT is also progressively acquired as a medium for gathering sensors data and this sensors data is used in medical, manufacturing, transportation etc.

Big data has huge prospects to enlarge and make utilization of smart city services. Big data is basically huge aggregates of data that can be examined by businesses to make suitable strategic moves and business decisions. Big data analysis is put into action to study large volumes of data to reveal patterns and get insights to extricate important information. Information and communications technology perform an important role in smart cities by making data available that are collected through information technology components [14]. The Internet of Things works by sharing information between connected devices while exchanging data that needs internet, wireless connections, and other means of communication. Smart cities utilize IoT devices mainly for fetching data and processing it efficiently to implement it in a specific area. Smart city sensors and connected devices gather data from different smart city gateways installed in a city and then examine it for better decision-making. With the use of Information and Communications Technology in smart cities environmental footprints will be reduced and result in the best utilization of resources. The sensors implanted in the city give a clear picture of what exactly is deficient in a city and how the present situation can be improved. Those areas which require improvement and upgrading can be identified by studying the present needs in a city through the effective use of data. Mapping infrastructure requirements in a city is not difficult when highly accurate data is used to pinpoint exactly where development is required. Transportation can be effortlessly controlled when big data comes into action. It is possible to restrain traffic through the proper use of historical data. The patterns can be studied through the analysis of data collected from transport authorities and it will result in reduced traffic congestion and help transport authorities come up with brilliant methods to manage and observe transport within the city. Big data analytics will also help in minimizing accidents. Big data plays an important role in smart cities by processing data collected through IoT devices so that further analysis can be made to acknowledge the patterns and requirements in the city.

2.6 Long Range and Narrowband-IoT

LoRa or Long Range is a spread spectrum modulation technique derived from chirp spread spectrum (CSS) technology and is the initial cheap implementation of chirp spread spectrum for business usage. LoRa uses license-free sub-gigahertz radio frequency bands like 433, 868 and 915 MHz. LoRa permits long-range transmissions of more than ten kilometer in rural areas with low power consumption. The technology is bestowed in 2 components, LoRa, the physical layer and LoRaWAN or Long-Range Wide Area Network, the upper layers. LoRaWAN is one of the protocols that was developed to outline the upper layers of the network. LoRaWAN is a cloud-based media access management layer protocol however acts in the main as a network layer protocol for managing communication between LPWAN gateways and end-node devices as a routing protocol. LoRaWAN is additionally accountable for managing the communication frequencies, data rate, and power for all devices.

Narrowband internet of Things or NB-IoT is a Low Power Wide Area Network radio technology customary developed by 3GPP to allow a large range of cellular devices and services. NB-IoT pay particular attention to indoor coverage, low cost, long battery life, and high connection density. NB-IoT uses a set of the LTE customary, however limits the bandwidth to one narrow-band of 200 kHz. It utilizes orthogonal frequency-division multiplexing modulation for downlink communication and Single-carrier FDMA for uplink communications [14].

2.7 Addressing

Due to the Internet, there is a phenomenal interaction among the people. Similarly, IoT offer interconnection of objects and things to provide smart environment. To this end, the capability of unambiguously identifying objects is crucial for favorable outcomes of the IoT applications. Uniquely addressing the large-scale combination of objects is important for controlling them via Internet. In addition to the mentioned uniqueness idea, reliability, scalability as well as persistence denote the key necessities to develop a unique addressing scheme [14].

2.8 *Blockchain Technology*

As we can see all around in research world, nowadays Blockchain technology is very hot topic in connected (Internet) world. As we know, the internet has already brought a lot of possibilities for smart cities to become more efficient, moving manual services to digital services and storing information paperless. The blockchain technology has provide secure distributed data, automated updated and trusted data accessibility between multiple parties in smart city use cases and Blockchain has works with multiple parties but best things are it's considered the privacy in connected system [15].

As in Smart Cities has mostly concentrate to provide services in efficient, secure, distributed, decentralized and trusted data to reduce the failure risk of system, lack of transparency and corruption. The Blockchain is provide the decentralized, distributed secure trusted data communication peer to peer network between multiple parties. These features are happened due to 4 pillars of Blockchain technology. These 4 pillar are shown in below Fig. 2 [15].

- **Ledger**: Ledger has provided the distributed data management in Blockchain network. It's shared the data with multiple parties. Ledger mainly overcome the high cost of Blockchain data storage.

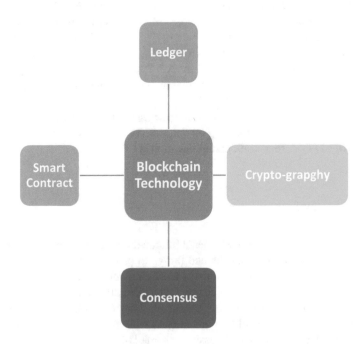

Fig. 2 Four basic blockchain pillars [15]

- **Cryptography**: Cryptography has encrypted the personal information and data in the network. Due to that, it has protected the privacy and secure the data in the networks.
- **Consensus**: Consensus has use for the data verification. It makes assure that transaction or messages are true and verified information. Consensus has made by the node who involved in network and they will get some reward based on their efforts.
- **Smart Contract**: The Blockchain is provides pre-defined rules to nodes before entering the network. It's helping to make trust and immutability feature in the networks. The nodes are involved in the execution of information in the network without any fear. They make the process faster and less expensive.

3 Smart City Applications

3.1 Automated Gardening System

One of the problems faced in daily life is a lack of time. In urban areas, people don't have time to look after their gardens. They forget to water their plants and because of which it becomes difficult for plants to stay healthy and alive. Therefore, it is vital to implement the automated system which will take charge of plants protection by considering all the different features of the home gardening system and also helps them to stay healthy. Plants are very advantageous for all human beings in many ways. Plants help in keeping the surroundings clean by cleaning air naturally and producing oxygen. Oxygen is the basic necessity of human beings. Some people love to have plants in their homes but due to less space or an inappropriate place for plants, they grow plants in pots on their balconies, but they don't have time to take care of the garden. Even we have less knowledge of how much water is needed by plants. Sometimes, we provide more amount of water than actually required by plants. Immoderate watering and less watering, both are unfavorable for plants. Plants require a specific amount of sunlight, water and temperature to stay healthy.

Using automated system, we can control these factors so that plants can grow in a better way. Automated garden helps people in maintaining their garden. The system will monitor the water intake of plants by measuring the water content of the soil. Type of soil is one of the important elements that have effect on plant growth. The water carrying capacity of the soil varies with the type of soil. We use a soil moisture sensor to measure the water content. Soil moisture sensors compute the volumetric water content in soil. Soil moisture sensors measure the volumetric water content indirectly by using some property of the soil such as electrical resistance, dielectric constant, or interaction with neutrons, as a proxy for the moisture content because measuring gravimetric water content is a tedious task to do and requires removing, drying, and weighting of a sample [16]. The soil moisture sensor is made up of two probes which are used to measure the volumetric content of water. The two probes permit the current to pass through the soil and then it gets the resistance value to

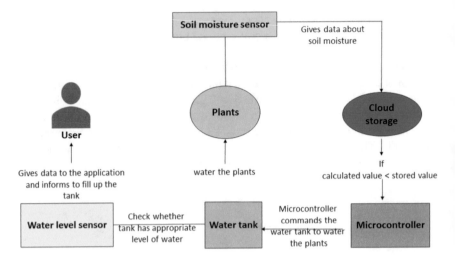

Fig. 3 Architecture of automated gardening system [17]

compute the moisture value. The soil will conduct more electricity when there is more water which means that there will be low resistance. Therefore, the moisture level will be higher. Dry soil is bad conductor of electricity, so when there will be small amount of water present in the soil, then the soil will conduct less electricity which means that there will be more resistance resulting in lowering the moisture level (Fig. 3).

If the value obtained is less than the default value stored in database, the microcontroller command to water the plants otherwise no action will be taken. Temperature influences the photosynthesis, respiration, flowering and many other factors in plants. Higher temperatures may result in the wilting of plants whereas lower temperatures may result in the poor growth of the plant. Therefore, to handle this kind of situation, the system has a temperature sensor. The sensor circuitry is sealed and thereby protecting it from oxidation and other processes and also has low self-heating. The light dependent resistor is a light controlled variable resistor. With increasing incident of light intensity, the resistance of a photoresist decreases, and it exhibits photoconductivity. A photoresist or can be used in light-sensitive detector circuits, light-activated and dark-activated switching circuits. In this automated system, this sensor is used to measure the intensity of light of the surrounding.

Water level sensor is used to measure the amount of water present in the tank which is used for watering the plants. The level measurement can be done either in continuous values or point values. The continuous level sensors measure level within a fixed range and discover the exact amount of substance in a definite place, whereas point-level sensors only specify whether the substance is above or below the sensing point. The point level sensors generally detect levels that are excessively high or low.

We use LCD screen to display the sensor readings. The whole system is connected to the Internet and user can also control the system with the help of internet.

3.2 Streetlight Automation

In modern cities, energy consumption is increasing day by day. Street lighting system is installed on the streets, which consumes a lot of electricity. It is on during daytime too when it is not even needed. Sometimes it happens that in some areas where there is very less passerby, the amount of consumption of energy is same as those areas where the frequency of passerby is high. LEDs are invented but yet in some areas old streetlamps are installed which consumes greater amount of energy as compared to LEDs. These old systems should be modified and automated so that energy consumption could be reduced and if any fault is detected, and then it also can be removed. Streetlight is very necessary to ensure the safety of vehicles and pedestrians. The streetlights in cities are using old techniques and it lacks latest technology. And because of which it is wasting electric power and also wasting manpower. The old streetlight uses sodium vapor lamps which emit huge amount of heat and gases in the atmosphere. In spite of which LED (Light Emitting Diode) lamps can be used. A light-emitting diode or LED is a semiconductor light source which produces light when current flows through it. The electrons in the semiconductor combine again with electron holes and release energy in the form of photons and this effect is known as electroluminescence. The LEDs have many benefits over traditional light sources such as lower energy consumption, longer lifetime, improved physical robustness, smaller size, and faster switching. The light emitting diodes are utilized in applications as diverse as aviation lighting, automotive headlamps, advertising, general lighting, traffic signals, camera flashes, medical devices etc. The LEDs lifetime is fifty times better than traditional lamps. They are both cost and energy efficient. They don't emit greenhouse gases which results in being eco-friendly to the environment. We can control the intensity of LED light using PWM (Pulse Width Modulation) technique which can help in light automation. Pulse Width Modulation or PWM is a technique for receiving analog outcomes with digital means. The digital control is utilized to generate a square wave which is a signal switched between on and off. This on-off pattern can produce voltages in between full on that is of 5 V and off that is of 0 V by altering the portion of the time the signal spends on versus the time that the signal spends off and the duration of on time is known as the pulse width. The pulse width will be altered or modified to receive varying analog values. If this on-off pattern is repeated fast enough with an LED, then the result is as if the signal is a steady voltage between 0 and 5 V controlling the brightness of the LED. The light will normally glow at less intensity maybe of 20% but when a sensor detects a vehicle, the light will automatically starts glowing at 100% intensity [4]. We can use LDR or Light Dependent Resistor to sense vehicles' presence. LDR is a light controlled variable resistor. The resistance of light dependent resistor decreases with increasing incident light intensity which means it shows photoconductivity. The light dependent resistor can be applied in light-sensitive detector circuits, light-activated and dark-activated switching circuits. The LDR is made up of a high resistance semiconductor. The LDR can have a resistance as high as several megohms in the

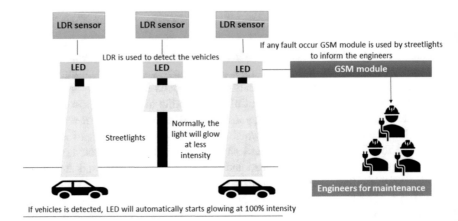

Fig. 4 The streetlight work process in smart cities

dark whereas it can have a resistance as low as few hundred ohms in the light [4]. The light emitted by vehicles' headlight is sensed by LDR and identify the existence of vehicles.

Global System for Mobile Communications or GSM is originally named as Groupe Special Mobile and is used to set up communication between a computing machine and a mobile device. A GSM/GPRS modem can perform operations such as receive, send or delete SMS messages in a SIM; read, add, search phonebook entries of the SIM; make, receive, or reject a voice call. GSM technology is used to monitor the status of streetlamps. The sensor installed in each streetlight if sense any fault, then it is notified to the base station using GSM module so it can be repaired as early as possible. The streetlight automation is a very efficient approach and can minimize energy consumption and provide more safety to vehicles and pedestrians. In Fig. 4 has shown the streetlight process. As we can see in figure, LDR detect the vehicles and transmits the signal and normally light will glow at less intensity after detected the vehicle and LED will glow 100% intensity. For maintenance, it will use the GSM module to broadcast the message to maintenance engineers.

3.3   Intelligent Transportation System

As the economies of countries are growing, people are buying more vehicles to travel. Some people are buying it for maintaining their social status while others are buying it as a necessity. And because of which the roads are full of vehicles. This leads to accidents which are very common nowadays. Technologies are advancing, new security measures are getting implemented to make the human's life more secure but when it comes to road accidents, we are still facing numerous cases of accidents and people are losing their precious lives. This gives encouragement to Intelligent Transportation system. Smart-Eye technology helps in making the transportation

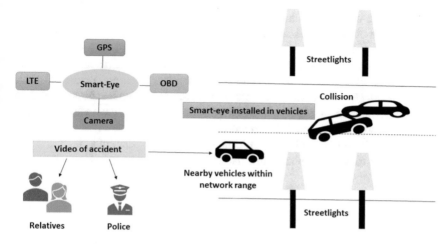

Fig. 5 SMART-EYE based intelligent transportation system

system smart by ensuring safe driving, preventing accident and monitoring of traffic [7] as shown in Fig. 5.

When an accident happens, there are chances of other vehicles' collisions and road rage which sometimes turn to very big problem and become difficult to handle. Smart-Eye is the solution for all these things. To prevent such accidents, smart eye can be installed in the vehicles and connected to vehicular networks. Smart-Eye has the ability to store location, driving video, and rear end collision both before and after the accident. In the event of an accident, it also stores the location of accident as a vital piece of information for the concerned authorities with the help of cloud networking.

Smart-Eye consists of LTE or Long-Term Evolution, GPS or Global Positioning System, OBD (On-Board Digenesis) and camera. Long-Term Evolution or LTE is a standard for wireless broadband communication for data terminals and mobiles connections which is based on the GSM/EDGE and UMTS/HSPA technologies. The main aim of LTE is to improve the efficiency and speed of wireless data networks by making use of new DSP or digital signal processing techniques and modulations that were developed around the thousands of years ago. And is also utilized to redesign and reduce complexity of the network architecture to an IP-based system with significantly lower the transfer latency compared to the 3G architecture. The LTE wireless interface is not compatible with 2G and 3G networks and thus it must be operated on a separate radio spectrum. Global Positioning System or GPS is a satellite-based radio navigation system possessed by the United States government and controlled by the United States Air Force. It is a global navigation satellite system and it provides time information and geo-location to a GPS receiver anywhere on or near the Earth where there is an unblocked view to four or more GPS satellites.

Smart-Eye can share real-time information among the vehicles connected to vehicular networks and secure cloud networks. It is used for prevention of accidents within a particular range. If an accident happens, the video of accident can be shared immediately to the other vehicles so that they can turn their directions and avoid traffic due to accident. It also shares the information of accidents with the location of accident to the concerned authorities and to the family members of the accident victim. Smart-Eye is used to store pre and post-accident data to be processed and at the same time transfer data to the end users who are nearby the ad hoc network without any mediation from third party electronics device so that they can receive the news about the nearby vehicles and the drivers can then decide for themselves whether to overtake the foregoing vehicles or not. Intelligent Transportation system is a very good solution to be implemented in smart cities [7].

3.4 Smart Garbage Automation

Garbage is a waste material that is disposed of by humans, generally due to lack of use. The main issue with our environment has been solid waste management which affects the health and environment of our society. The detection, observation and management of wastes are one of the major problems of the present time. The traditional way of monitoring the wastes in waste bins manually is a complicated process and uses more human effort, time and cost which can easily be reduced with our present technologies. The major issue in the waste management is that the garbage bin at public places gets cram-full with garbage in advance before the beginning of the next cleaning process and which in turn leads to different hazards such as bad smell and un-cleanliness of that area which is the root cause for spread of different diseases. Garbage system should be automated to prevent these hazardous situations and maintain public cleanliness and health [6]. A smart garbage system consists of an ultrasonic sensor placed in a dustbin, which is used to measure the levels of garbage in the dustbin. Ultrasonic sensors are a kind of acoustic sensor which is divided into three broad categories such as transmitters, receivers and transceivers. Transmitters convert electrical signals into ultrasound, receivers convert ultrasound into electrical signals, and transceivers can both transmit and receive ultrasound. Arduino is used as a microcontroller in this design. Arduino is an open-source hardware and Software Company that designs and manufactures single-board microcontrollers and microcontroller kits for building digital devices and interactive objects that can sense and control both physically and digitally. Whenever the garbage in the dustbin reaches up to the certain level which is not appropriate or after that overflow can occur, the information is sent to the server node with the help of Bluetooth (Fig. 6).

Bluetooth is a wireless technology for interchanging data between fixed and mobile devices over short distances using short-wavelength ultra-high frequency radio waves in the industrial, scientific and medical radio bands, from 2.400 to 2.485 GHz and creating personal area networks. The server node then alerts the user and concerned authorities about the pre-overflow condition to the user by sending an SMS using GSM module. MQ4 sensor is also used to sense the hazardous gases released by the garbage and alert the user by sending the message and inform

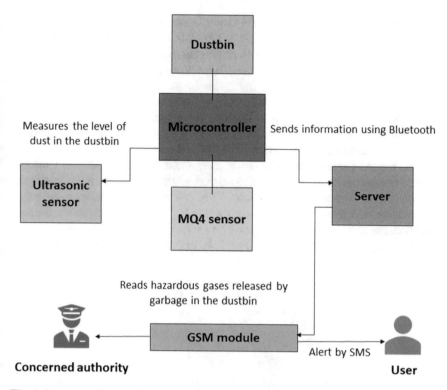

Fig. 6 Overview of smart garbage automation in smart cities

them to take precautionary measures. MQ4 gas sensors are used for detecting gases such as methane and natural gas and have high sensitivity to methane and natural gas. The resistance of the sensor differs depending on the amount of gas [17].

This system is very useful for the society as it is cost effective, time saving and uses wireless sensor network. If it is used for all dustbins placed in smart cities, the condition of smart cities can be improved, and dustbins can be properly monitored. This way overflowing of the dustbins and generation of hazardous gases can be avoided which makes people life more secure. This idea helps in the reduction of solid waste management problems and helpful in increasing public health safety precautions by Government.

3.5 Blockchain Based Autonomous Vehicle Communication

As vehicles are more autonomous and smart cities are more connected by using more communication tools such as sensors, instruments etc. For autonomous vehicles, smart city must be using more IT infrastructure to store, protect and analyze the data generated by autonomous vehicles in networks. Autonomous vehicles have improved their performance by combine and evaluating data available from smart cities. As much as vehicles will improved in their autonomous feature that much data has generate and available in the networks. Those generated data need protection and also need to trust those data's are big issues in smart cities, In here, Blockchain technology can help to protect and make assure that existing data's are verified and authenticated that can help to improve the vehicle communication and safe travelling in the smart cities. Blockchain technology to support trusted environment for autonomous vehicles (AV) communication in Smart Cities [15].

This trusted environment provides a secure distributed decentralized mechanism to communicate without sharing personal information of AV in an Intelligent Transportation System (ITS) of Smart Cities. We can see the Blockchain based autonomous vehicle communication networks in Fig. 7. In Fig. 7 has shown the how can Blockchain technology will improve and secure the vehicle communication in the smart city's region. As shown in Fig. 7 we have categories 2 Blockchain server

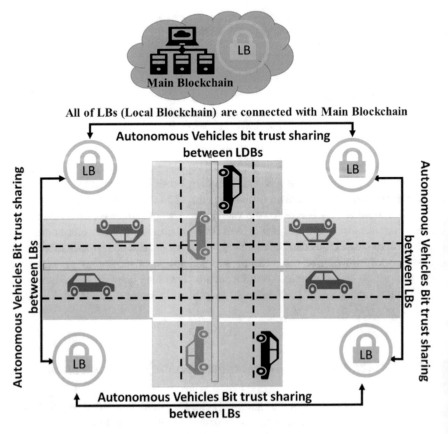

Fig. 7 Blockchain based autonomous vehicle communication networks

such as Local Blockchain (LB) and Main Blockchain (MB) with a secure and unique crypto ID, called as Trust Bit. This Trust Bit ensures trustworthiness' among vehicles. Vehicles use and verify the Trust Bit with the LB to communicate with other vehicles. Therefore, this method has been found to be trusted based on MB, LB and Trust Bit to check vehicles trustworthiness. The Blockchain technology with vehicular networks concentrates on secure and fast communication between intelligent vehicles (internet enabled self-driving cars. LB is provide the peer to peer network communication between vehicles and it has over ridable server and MB will store the data securely for long time. LB has store the information only present time and it will override the data once server has full and MB will store the event-based data lifelong with encryption form [16].

The Blockchain is used to remove the centralize authority for communication among AVs in smart cities. Therefore, we have considered a crypto Trust Bit that shall help to improve the privacy of AV. Trust bit provide fast and secure communication between AVs in Smart Cities. It also helps to detect the history of AV in the different region of Smart cities. For AV communication data storage, we have proposed two Blockchain mechanisms, LB and MB. The LB has limited memory and computational power. It ensures trustworthiness of Vehicles directed in its region. All LBs are connected with each other sharing the information in circular fashion and overwriting its database at periodic intervals. Contrary, MB is responsible for storing the complete ITS data and updating its database from LDB when there is no transaction between the LDBs. MB is responsible for providing the complete history of any vehicle ubiquitously in Smart Cities [18].

4 Challenges

4.1 Security and Privacy

Smart cities are anticipated to improve the quality of daily life, encourage sustainable development, and improve the functionality of urban systems. Many smarts have been implemented still there are security and privacy issue which is major concern that require countermeasure. However, traditional cyber security protection strategies cannot be exerted directly to these intelligent applications because of the heterogeneity, scalability, and dynamic characteristics of smart cities. It is vital to be aware of security and privacy threats while designing and implementing new mechanisms or systems. Authentication is a basic need for different layers of a smart system and is required to prove identities and ensure that only authorized clients can access services across a heterogeneous system.

4.2 Reliability

There are some reliability issues that have arisen in the IoT based system [19]. The communication with vehicles is unreliable because of the mobility of vehicles. And also, the presence of various smart devices will cause some reliability challenges in terms of their failure.

4.3 Large Scale of Information

Some specific situations require the interactions between large numbers of embedded devices which are probably distributed over wide area environments. The IoT systems give a suitable platform to analyze and integrate data coming from different devices. However, such large scale of information collected at high rate requires appropriate storage and computational capability which makes usual challenges difficult to conquer. In addition to this, the distribution of the IoT devices can affect the monitoring tasks because these devices must handle the delay related to dynamics and connectivity.

4.4 WSN Communication

Sensor networks can be considered as one of the most significant technologies to enable the Internet of Things. This technology is able to mold the world by providing the potential of measuring and understanding environmental indicators. Recent developments in technologies have provided devices with high efficiency and low-cost to engage remote sensing applications in large-scale. In addition, smart phones are connected with a variety of sensors and as a result they enable a diversity of mobile applications in several areas of IoT. Due to this, the major challenging task is to process the large-scale data of the sensors in terms of energy and network limits and various uncertainties.

5 Conclusions

In this chapter, smart city and technologies prevailing in smart cities have been discussed. Smart city is an urban area, which uses IoT devices and sensors to implement smart application in cities. The motivation for smart city is the lack of time, reduction in human efforts. IoT is playing a major role in the development of smart cities. Some of the smart city applications are discussed above such as smart garbage automation, automated gardening system, intelligent transportation system and smart streetlight automation which will reduce the human efforts provide more security and safety to human beings. The widespread use of smart city technologies has caused many security and privacy issues. We have discussed many challenges that have emerged in smart city technologies. Although various protection mechanism and strategies have been developed in recent years still there is need for a better solution either a framework or a protection models in both industry and academic fields. Various protection mechanisms and strategies have been developed in recent years. However, there is a long way to go to satisfy the multiple security requirements of these rapidly developing smart applications. It is reasonable to predict that in the following few years, mitigating the presented challenges will be the primary task of smart city-related studies.

Acknowledgements This wok was a part of summer internship program supported by MtoV Inc. Therefore, Author would like to Thanks Dr. Dhananjay Singh for his guidance and support to work on this chapter.

References

1. Hammi B, Khatoun R, Zeadally S, Fayad A, Khoukhi L (2017) Internet of things (IoT) technologies for smart cities. IET Netw 7. https://doi.org/10.1049/iet-net.2017.0163
2. Serrano W (2018) Digital systems in smart city and infrastructure: digital as a service. MDPI: Smart Cities 1(1):134–154. https://doi.org/10.3390/smartcities1010008
3. Schaffers H, Komninos N, Pallot M (2012) Smart cities as innovation ecosystems sustained by the future internet. FIREBALL White Paper, EU, pp 1–65
4. Revathy M, Ramya S, Sathiyavathi R, Bharathi B, Maria Anu V (2017) Automation of street light for smart city. In: International conference on Communication and Signal Processing, India, 6–8 Apr 2017
5. Singh D, Singh M, Singh I, Lee H-J (2015) Secure and reliable cloud networks for smart transportation services. In: The 17th IEEE international conference on Advanced Communication Technology, (ICACT2015), Phonix Park, South Korea, 1–3 July 2015
6. Esmaeilian B, Wang B, Lewis K, Duarte F, Ratti C, Behdad S (2018) The future of waste management in smart and sustainable cities: a review and concept paper. Waste Manag 81: 177–195
7. http://k-smartcity.kr/english/smartcity/smartcity.php
8. Alberti AM, Singh D (2013) Internet of things: perspectives, challenges and opportunities. In: International workshop on telecommunications (IWT 2013), INATEL, Santa Rita do Sapucai, 6–9 May 2013, pp 1–6
9. Singh D (2015) Secure 6lowpan networks for e-healthcare monitoring applications. J Theor Appl Inf Technol 76(2):143–151
10. Zhang J, Tian GY, Marindra AMJ, Sunny AI, Zhao AB (2017) A review of passive RFID tag antenna-based sensors and systems for structural health monitoring applications. Sensors 17:265
11. Singh D, Singh M (2015) Internet of vehicles for smart and safe driving. In: 2015 International conference on Connected Vehicles and Expo (ICCVE), Shenzhen, China, 19–23 Oct 2015
12. Khan I, Singh D (2018) Efficient compressive sensing based sparse channel estimation for 5G massive MIMO systems. AEU—Int J Electron Commun 89:181–190. https://doi.org/10.1016/j.aeue.2018.03.038 (Elsevier)

13. Singh I, Lee SW (2018) Comparative requirements analysis for the feasibility of blockchain for secure cloud. In: Kamalrudin M, Ahmad S, Ikram N (eds) Requirements engineering for internet of things, APRES 2017. Communications in computer and information science, vol 809. Springer, Singapore

14. Singh D (2013) Developing an architecture: scalability, mobility, control, and isolation on future internet services. In: Second international conference on Advances in Computing, Communications and Informatics (ICACCI-2013), Mysore, India, 22–25 Aug 2013

15. Singh M, Kim S (2017) Introduce reward based intelligent vehicle communication using blockchain technology. In: 14th International SoC Design Conference (ISOCC 2017), Grand Hilton Hotel, Seoul, South Korea, 5–8 Nov 2017

16. Singh M, Kim S (2018) Trust bit: reward-based intelligent vehicle commination using blockchain paper 2018 IEEE 4th World Forum on Internet of Things (WF-IoT), Singapore, pp 62–67. https://doi.org/10.1109/wf-iot.2018.8355227

17. Abdel-Shafy HI, Mansour MSM (2018) Solid waste issue: sources, composition, disposal, recycling, and valorization. Egypt J Petrol 27(4):1275–1290. ISSN 1110-0621. https://doi.org/10.1016/j.ejpe.2018.07.003

18. Singh M, Kim S (2018) Branch based blockchain technology in intelligent vehicle. Comput Netw. https://doi.org/10.1016/j.comnet.2018.08.016

19. Albreem MAM, El-Saleh AA, Isa M, Salah W, Jusoh M, Azizan MM, Ali A (2017) Green internet of things (IoT): an overview. In: IEEE international conference on Smart Instrumentation, Measurement and Applications (ICSIMA), Songkla, Thailand

An Architecture for e-Health Recommender Systems Based on Similarity of Patients' Symptoms

Valerio Frittelli⬡ **and Mario José Diván**⬡

Abstract Nowadays, data are generated both by users and other systems deriving new data from the previous ones for supporting decision making. The Electronic Health Records contains from structured data (e.g. hospital id, etc.), semi-structured data (e.g. a Health Level Seven-based records), to unstructured data (e.g. patient's symptoms). The big challenge with health in smart cities is associated with the prevention, both the business and human health point of view. That is to say, avoid the propagation of certain diseases' patterns is the best option no just for people, but also from the city's health and the local economy. Thus, an architecture able to integrate into an Organizational Memory the medical data coming from heterogeneous repositories with the aim of gathering different kinds of symptoms is introduced. The query in the architecture is understood such as an unstructured text (i.e. symptoms) or an electronic health record. In this sense, the architecture is able to reach similar cases from the organizational memory based on a textual similarity analysis for limiting the search space. Next, using the International Classification of Diseases is possible to convert a case to a vector model representation in order to compute metric distances and get other cases order by a level of similarity. Each query answer contains a set of recommendations based on the frequency of diagnoses related to similar cases are given in order to share previous experiences. The processes point of view related to architecture is outlined. Finally, some conclusions and future works are outlined.

Keywords Medical health record · Organizational memory · Similarity · Unstructured data · Metric spaces · Decision making

V. Frittelli
National Technological University, Maestro López and Cruz Roja Argentina, 5000 Córdoba, Argentina
e-mail: vfrittelli@frc.utn.edu.ar

M. J. Diván (✉)
National University of La Pampa, C. Gil 353. 1st Floor, 6300 Santa Rosa, Argentina
e-mail: mjdivan@eco.unlpam.edu.ar

© Springer Nature Singapore Pte Ltd. 2020
D. Singh and N. S. Rajput (eds.), *Blockchain Technology for Smart Cities*,
Blockchain Technologies, https://doi.org/10.1007/978-981-15-2205-5_8

155

1 Introduction

The volume of data related to the people's health is increased each day through the interaction between patient and doctor but also from interaction among medical systems. These kinds of medical systems commonly have a specific application, giving a determined point of view in the medical analysis, collaborating with other systems in order to satisfy a superior aim, for example, the patient monitoring [1, 2].

The medical systems have an important data heterogeneity which is common and even expected in them. In this way, the medical analysis could imply from a bone scanning, image, audio, video, to diagnosis expressed such as an unstructured text, among others [3].

Taking the notion of continuous growing associated with each type of medical data sources and considering its level of heterogeneity, the medical decision-making supporting becomes in a complex activity. That is to say, it is necessary to deal with a huge data volume but with the data heterogeneity too. However, the heterogeneity of data refers to the syntactical point of view (i.e. the data structure represented by an image, audio, etc.,) but also to the semantic point of view (i.e. what is the meaning which is represented in each record) [4].

The Clinical Information Systems (CIS) are oriented to implement procedures and work's schemas necessary to automatize the patient attention, its monitoring jointly with the clinical records. Each time in which a patient meets with their doctor, some diagnoses, appreciation or observation is incorporated into their medical track. That kind of diagnoses, explanation of symptoms, observations, among others are documented by the expert in a free narrative way. That is to say, the expert expresses in their own words and under a technical narrative such appreciations keeping record in the system [5].

Because a person can live in different places throughout his life, or even be attended in different health centres, the medical records naturally will be distributed. Thus, it is highly possible that they are heterogeneous in terms of their structure. In addition, each medical expert writes the symptoms and diagnoses using its own style and words, which will imply an additional complexity in the posterior analysis [6].

In this sense, the integration of different records in a homogeneous way in order to virtually centralize the data set for the analysis of symptoms and diagnoses based on unstructured text implies the development of a huge number of pre-processing tasks. Even the possibility of tagging each record in order to indicate some profile or pattern related to some illness would be very useful for recommender systems [7].

The text similarity functions are especially interesting in the medical records when symptoms, diagnosis and tagging are incorporated because allow identifying similar situations, be it associated with the same patient or not. Thus, different early actions could be taken in order to prevent some illness or the propagation of it. For example, if a doctor is attending patients in a Hospital, and the system is able to identify various patients with similar symptoms using text similarity on medical records, it will allow anticipating a risk situation. That is to say, supposing that one medical record is identified such as cholera, then all the similar cases of other patients could be associated with this illness, which implies that there is a possibility to reach an early recommendation on patients, carrying out preventive actions in order to avoid the propagation of the illness [8].

In this kind of situations where similarity based on the unstructured text should be computed, the exact matching is not an alternative for answering queries. That is to say, when a doctor attends a patient, the system does not find a record perfectly matching with symptoms and diagnoses, but similar situations which eventually could be related to a given situation or not. Depending on the level of similarity between the two situations, the associated likelihood will be upper or lower. Thus, the search strategy is oriented to a metric space in where each medical case could be represented as a point, while the system will find the nearest cases considering the case of the query as a virtual centre. In other words, the expected behaviour in the search strategy is similar to the centre choosing in a clustering algorithm [9].

The Organizational Memory (OM) allows structuring a set of knowledge and previous experiences in an organization, using a specific organization which allows querying cases, and even, reasoning for obtaining new knowledge based on the previous ones [10–12]. The OM could be based on ontologies for sharing the common concepts but also based on specific domain ontologies for extending and specializing the knowledge [13].

The recommender systems allow articulating the search strategy based on text similarity with organizational memory. On the one hand, given a medical case, the search strategy will find similar cases in a repository, analysing symptoms, diagnoses and associated tags from unstructured data. On the other hand, once the profile or pattern of cases has been determined, the organizational memory provides the necessary knowledge and previous experiences for deriving actions and recommendations [14].

As the main contribution, a Patient Symptoms-aware Architecture based on an Organizational Memory (PSAbOM) is introduced. PSAbOM incorporates an organizational memory which organizes the medical records describing symptoms, diagnoses and tags (when they are known) as cases. In addition, it supports a Similarity-based text searching which allows obtaining similar medical cases in order to prevent medical risk situations and learn from previous experiences. Thus, given a case, the architecture contains a recommender which is able to reach the likely actions based on the pertinence and analysis of frequencies from pre-existent situations (i.e. other cases).

The chapter is organized into seven sections. Section 2 introduces some related works, describing the common points jointly with their associated differences. Section 3 synthetically describes the textual similarity analysis strategy. Section 4 outlines the organizational memory, describing the perspectives, cases organization, similarity functions, and search strategy. Section 5 introduces the architecture outlining contextual limitations, data collecting strategy, data pre-processing, the processing itself jointly with the recommendation perspective. Section 6 synthetically schematize the architecture through an application case. Finally, some conclusions and future works are introduced.

2 Related Works

The related works associated with health could be organized in relation to the text similarity analysis, Case-based reasoning, recommender systems, and Big Data repositories.

From the Text-similarity analysis point of view, Kenter and Rijke [15] propose to represent short texts under the way of vectors, using word embeddings for giving a semantic matching. The semantic matching means to establish a relationship between a given word with some kind of meaning space. In this way, as nearer two vectors are in the space as a closer would be the meaning between them.

Kashyap et al. [16] describe SemSim system which combines semantic analysis, combined with data mining algorithms and external linguistic resources in order to compute the level of semantic equivalence. The proposal is not limited to short texts, because of that the system is able to process and analyze long texts for determining how nearest the meaning of two given documents are.

Mrabet et al. [17] introduce a new measure named TextFlow, which considers the sequential aspect of a language in the determination of a given meaning. In this sense, a new representation based on curves is introduced, in which the current word represents the origin, while the next words in a sentence are associated with the displacement on the curve. Thus, from the origin and such displacement, the new similarity measure is computed.

Yao et al. [18] introduce an algorithm-based long short-term memory (LSTM) encoder, which is able to detect semantic similarity from words in a supervised way. The evaluation of short words finally is made through cosine distance in order to determine the level of similarity based on a trained neural network.

As a similitude in terms of text analysis, in our architecture, the similarity analysis is based on a vector model in order to determine the similarity between sentences expressed as vectors based on the distance between them. However, the proximity between sentences is the input for the recommender system in order to reach similar situations. That is to say, the underlying idea is to reach medical similar cases for getting feedback based on previous experiences.

From the Case-based reasoning and recommender systems, Miotto and Weng [19] proposed a case-based reasoning framework oriented to identify similar patients through the cosine distance. Each medical record is analyzed and compared among patients, using the cosine distance in order to determine whether a patient is eligible for researching or not. The precision of this approaching was analyzed through a binary regression, taking the cases' similarity jointly with the previous knowledge on eligibility or not of a given patient based on medical records. As a coincident point, the architecture looks for similarity based on a vector model, while as a difference, our proposal incorporates the recommending based on previous experiences.

Gómez-Vallejo et al. [20] introduce a cased-based reasoning system focused on the surveillance and diagnosis of infections. The authors gather and integrate data coming from heterogeneous and unstructured medical data sources. Next, they use data mining techniques to extract evidence from them in order to deduce new knowledge. As a difference, our proposal incorporates the recommending as feedback derived from the obtaining of new knowledge.

Su et al. [21] introduces a strategy based on a medical ontology for describing the knowledge base. From that knowledge, a case-based reasoning schema is used in order to risk prevention and projection. A case focused on type II of diabetes mellitus is introduced. As similitude, the recommending is considered as a component of the strategy, but as a difference, the proposal starts from data organized in terms of an ontology and not unstructured text as in our case.

The Electronic Health Records (HER) in their different formats complement the data available in big data repositories, which they are characterized and contextualized by the heterogeneity, volume and velocity in which data is incorporated [22]. Our proposal considers from the unstructured data pre-processing related to symptoms and diagnoses expressed as cases, to the recommendation strategy oriented to capitalize on the previous experiences and knowledge in order to prevent medical risks. The architecture introduces an integrated vision from the narrative style associated with the medical diagnoses, to the prevention schema based on recommendations which are especially useful in the context of smart cities.

3 The Textual Similarity Analysis Strategy

The searching, retrieving, and ordering based on the relevance of unstructured text documents, which contents are coincident with a given pattern entered by a user is typically an area from Information Retrieval (IR). Basically, it is the challenge that it needs to be solved by searching engines, independently they are looking for contents on the web or in a local context (e.g. a local file system). Different kind of mathematical models and alternatives were proposed for designing this kind of applications [23]. The underlying idea is to approach the similarity between two documents and order the results based on the level of relevance with the entered query, for that aim boolean, vector, statistic, probabilistic or semantic approaches could be used [24]. Typically, a searching engine uses a strategy or combine a couple of them, presenting the final result as a list to users, ordering results by relevance in terms of a given query.

The fact of measuring the unstructured content is really a challenge because the content could be written in an informal way, including abbreviatures, specific language, regionalities, or specific conventions/expressions related to a given field. This is precisely the case of the human health and their associated electronic records, in which the doctors write the patient's symptoms as text using a colloquial way, giving recommendations and treatments in the same way. Even when the electronic health records could have a set of structured information (e.g. the patient ID, visit's date, etc.), certain searches need to compare those text from the doctor for establishing or looking for some kind of common pattern.

The kind of models applied in IR traditionally could be classified into different groups in relation to the used strategy [25–27], as follows:

1. *Boolean-based Models (or Theoretical Set Models)*: The most elemental model is known as a Boolean standard model, in which the document relevance given a query is measured in a binary way. That is to say, the relevance value is defined whether the document contains or not the searched words. Thus, the relevance of 1 indicates that the searched words are contained in the document, while 0 indicate the absence of them in the document [27]. The basic model is not able to get a relevance ordering in the results given a query, with that aim the Extended Boolean Model was introduced, which is able of assigning specific weightings, measuring the partial matching with the entered query [28]. Alternatively, the Fuzzy Logic Model allows incorporating the idea of belonging to a set, being able to measure the level of partial belonging [29].

2. *Vector-based Models (or Algebraic Models)*: the most elemental model in this family, known as Standard Vector Model, represents each document as a vector containing as many dimensions as words are contained in the vocabulary. Each coordinate measures the weighting or value which each term has in relation to the represented document. Next, the distance between vectors is calculated using the angle's cosine between them, which is known as cosine distance. Other models under this family, such as the Generalized Vector Space Model or Latent Semantic Indexing Model, introduce the semantic analysis which is not considered in the standard model [26, 30, 31].

3. *Probabilistic Models*: the most elementary model, known as Standard Probabilistic Model, tries to estimate the likelihood that a user retrieves a document "*d*" which is most relevant in a query "q". Alternatively, other models such as the Belief Network Model and Inference Network Model based on the concept of Bayesian network used for computing the joint occurrence were introduced [32–34].

4. *Interactivity Models*: This family is oriented to study the user and their associated actions through the query process [25]. Under this category could be found cognitive models [35], User behavior-based models [36], and Popularity Ranking Models (or simply PageRank) [37].

5. *Artificial Intelligence Models*: Under this family, it is possible to find models in which the knowledge or previous experiences are used for deriving or reasoning new pieces of knowledge in order to decrease the level of uncertainty. For example, models such as Logical Models [38, 39], Knowledge-based Models [40], Neural Networks [41], Evolutive Algorithms [42–44], Natural Language Processing [45, 46], Ontology-based Knowledge Representation [47], Machine Learning [48], among others.

The mentioned models use some kind of indexing based on the available data on the documental base for searching and retrieving a subset of documents suitable in relation to the given query. The indexing structure could include a vocabulary containing all the terms jointly with a list describing the documents in which they have been located (it is known as a posting list). Both the vocabulary and the list are known as inverted indexing structure [26]. Thus, given a query, the retrieved documents are ordered by the relevance and the user will choose from the ordered list those documents based in its own interest. In addition, there are proposals oriented to give support to users at the moment in which they need to choose documents based on its interest, but in such cases, it is useful a new indexing based on similarity. Thus, alternatives such as Similarity Searching based on Metric Spaces [49], using Nearest-Neighbor Searching, or Range Searching are introduced [37].

4 Organizational Memory Based on Cases' Similarity

This section introduces the analysis of cases' similarity in the context of an organizational memory oriented to electronic health records. For better understanding, it is organized into four subjects. Initially, a global perspective of the underlying idea and concepts are given. Next, a description of the cases organization into organizational memory is presented. Followed, the application perspective of the similarity functions is described. Finally, the search strategy is outlined.

4.1 Perspectives

The Electronic Health Records (EHR) related patients in health centers incorporate both well-structured data (e.g. patient's information, social insurance ID, etc.) and a lot of unstructured textual data (e.g. a medical observation, an informal recommendation, a suggested diet, etc.) [50].

Depending on the kind of searching for clinical records, the kind of search strategy to be chosen could vary. On the one hand, when the search strategy is looking for specific antecedents (e.g. symptoms, tracks, etc.), a structured search is able to be carried out using a Structured Query Language (SQL). On the other hand, when the information retrieval needs to dive on a huge of unstructured medical text, it will require the computing of the level of similarity even knowing that it is not as precise as which SQL is. However, the last alternative allows arriving at similar situations answering a given query by means of proximity to the searched concept, which will require string analysis, text pattern analysis, natural language processing, among other techniques.

The International Classification of Diseases (ICD) is a taxonomy in which illnesses, symptoms, and other aspects associated with the human health are coded in order to make easier the understandability of the medical information in medical records. The taxonomy is published by the World Health Organization (WHO), and currently, version 10 (it is known as ICD10) is valid from 1990. However, it is planned that version 11 replace version 10 in 2022[1].

---

[1]https://www.who.int/classifications/icd/en/.

The ICD10 taxonomy assigns alphanumerical codes to each identified illness, but also death causes, symptoms, and other factors related to the health. The codes are divided into twenty-two chapters, in which each one has associated with some set or kind of general issue able to affect the health. Each chapter is break-down in blocks which allow a higher level of detail in relation to each disease. The max length of an ICD10 code is six characters under the format X00.00, where "X" represents a letter while the "0" are numbers. The point is used for separating the chapter/block ID of the specific associated symptoms. Thus, X00 identifies a chapter and block, while the numbers located next to the point are used specific health factors. For example, the code H70 is related to "Mastoiditis and related conditions", while the code H70.1 is specifically used for describing "Chronic Mastoiditis".

The ICD10 taxonomy tolerates around 14,000 different codes, but if the adapted versions jointly with sub-taxonomies are incorporated, the volume of codes is increased up to 70,000 different codes. It could be used in different ways for storing, retrieving, analyzing, sharing, and comparing data related to health records [51]. The codification each electronic health record using ICD10 is a complex activity because of the volume of codes and the possibility of using more than one in each case. However, there are works oriented to develop techniques for making easier the coding through natural processing techniques based on neural networks [52, 53].

4.2 Cases Organization

The cases into the organizational memory contains the minimal following information:

a. **ID**: An ID identifying the patient for chronology purpose. It allows establishing a timeline in relation to different attentions and diagnoses for a given patient.
b. **Timestamp**: It indicates a timestamp in which date, time and time zone are detailed.
c. **Symptoms**: A narrative description of the set of symptoms manifested by a patient is incorporated. This information is entered by the doctor based on the analysis, examination and the conversation with patients.
d. **Narrative Diagnostic**: A narrative explanation related to the professional diagnostic based on the incorporated symptoms. The diagnostic is incorporated by the doctor as an additional detail of the ICD10 codes.
e. **Coded Diagnostic**: A comma separated list containing ICD10 codes applicable for the diagnostic.
f. **Treatment**: A set of recommendations incorporated in a narrative way related to the diagnostic.
g. **Risk Profile**: A level of risk expressing 1 as no risk, while 5 indicates an elevated risk. This field is incorporated by the doctor expressing the level of severity for the given symptoms. The complete ordinal scale indicates:

(1) No Risk
(2) Minor Risk
(3) Moderate Risk
(4) Upper Mean Risk
(5) Elevated Risk.

Sadly, the missing value is not a new thing, this kind of challenges is common along with the integration of different kinds of datasets, and the electronic health records are not the exception [54, 55]. In the Organizational Memory, the only fields with warrantied value for each case are ID, timestamp, symptoms, and risk profile.

The risk profile is indicated by the doctor, and it expresses a code with a level of risk for the patient's health based on the symptoms' analysis. It acts as a complement of the diagnoses, and it is especially useful when the diagnostic is not determined or missing, mainly thinking in give some kind of orientation about the illness or the severity of the pathology.

The narrative diagnostic jointly with the ICD10 codes and the treatment could be missing because it is possible that the professional (i.e. the doctor) had not arrived at an adequate conclusion which allows identifying a pathology.

4.3 Application of Similarity Functions

Once the unstructured textual data is incorporated under the form of diagnoses or symptoms, the critical point is to determine the level of similarity between them given a query. This challenge is known as similarity-based searching or searching for proximity. The underlying idea is to measure the distance between two objects "a" and "b" which belong to a "C" set. An initial approaching would indicate that given the two objects, there exist some function which allows returning a number representing the distance, that is to say, a level of similarity or dissimilarity between them. When the function simultaneously satisfies the properties of Non-negativity, symmetry, reflexivity, strict positivity, and triangular inequality is known as a *metric distance function*. In addition, the set in which the function is applied is known as *metric space* [37].

Given two texts "a" and "b" which belong to a "C" set, it is possible to define a function "d" as d(a, b) in order to obtain the distance between them. Thus, $d:CxC \rightarrow \mathbb{R}$ is considered a metric distance function if and only if the mentioned properties outlined in Table 1 are simultaneously satisfied [49].

The non-negativity property establishes that the distance between two objects always will be positive or equal to zero. The property of symmetry talks about the commutativity of the distance between two points. In other words, the distance from the point "a" to "b" is equal to the distance from the point "b" to "a". The reflexivity indicates that the distance of an object to itself is always zero.

The property of strict positivity defines that the distance between two different points always will be positive (i.e. upper than zero). However, this property could be

Table 1 Required properties for a metric distance function

Property	Description
Non-negativity	$\forall a, b \in C : d(a, b) \geq 0$
Symmetry	$\forall a, b \in C : d(a, b) = d(b, a)$
Reflexivity	$\forall a \in C : d(a, a) = 0$
Strict positivity	$\forall a, b \in C \wedge a \neq b : d(a, b) > 0$
Triangular inequality	$\forall a, b, c \in C : d(a, b) \leq d(a, c) + d(c, b)$

not always satisfied because a distance between two different objects could be zero. Thus, when all the mentioned properties are satisfied with the exception of strict positivity, the distance function receives the name of pseudo-metric distance.

The triangular inequality is a key asset in the searching engines who wish to reach similar objects in a "C" set given a query "x". Geometrically, the underlying idea indicates that when it is necessary to measure a distance between "a" and "b" points, it is possible to find a point "c" in which the distance between "a" and "b" will be lesser or equal to the distance from "a" to "c" plus the distance from "c" to "b" [37, 49].

The challenge is to outline an algorithm or analytical function which allow defining a distance function given a "C" set. In this sense, an approaching indicates that it is possible to represent each object belonging to a "C" set as a vector. In other words, the objects (e.g. documents) try to be represented as a vector in which each dimension is related to a given object's characteristic. The number of dimensions to be used in a vector representation will depend on the number of wished characteristics. Thus, given two documents represented as vectors, it is possible to use a distance function for measuring the distance between them. When closer to zero is the measured distance between them, more similar will be the objects.

In this way, it is possible to think about an electronic health record or organizational memory record represented as a vector with the wished dimensions. For example, each ICD10 code could be represented as a dimension in order to characterize the vector. Even when ICD10 could reach around 14,000 different codes, a health record normally could be associated with six or seven codes [52]. Thus, when the ICD10 code is applicable in a health record, the corresponding position in the vector could be indicated with 1, while the absence could be indicated with a zero.

Thus, if the "C" set contains all the vector representations in k-dimensions (in. where k is the number of characteristics to be represented) of the health records, then "C" would be a Euclidean vector space because it is possible to apply the Euclidean norm of each vector (which is coincident with the longitude or module of each vector). For that reason, the "C" space is known such as a *normed vector space*.

A normed vector space is a metric space because the norm induces the distance function. That is to say, given a normed vector space called "C", it is possible to define the distance "d" between two vectors ("a" and "b") as the norm, module or longitude of the difference vector between "a" and "b" such as is shown in Eq. 1:

$$d(a, b) = \|a - b\| \tag{1}$$

In this way, considering the electronic health records represented as vectors, which integrate a set named "C", it could be possible to reach similar health records using a distance function. Thus, the definition of a metric distance function on a "C" set represents a key asset for analyzing the similarities among their elements. In addition, the organization of each element in the "C" set represents a smart decision to boost up the effectivity associated with the queries' answering on big data repositories. Different data organizations have been proposed, but all of them share a common aspect, the searching algorithm should belong to a sublinear order in the worst case (i.e. $O(n^{\alpha})$ in where $0 < \alpha < 1$). In other words, a searching algorithm should not read each document contained in the big data repository, it should be able to avoid an interesting number of readings for optimizing the answers' time [56, 57].

However, even when the metric distance function is defined, it is important to highlight that the searching requirements could be different depending on the context. In general, three kinds of the searching process could be outlined:

1. *Range Query*: it is responsible for retrieving all the items belonging to the "C" set which are located within an "r" distance from an "x" searched item.
2. *Nearest Neighbor Query*: it retrieves the object in the "C" set which is nearest from the "x" searched item.
3. *K-Nearest Neighbor Query*: it retrieves a set of "k" objects which are the most similar to the "x" searched item.

There are a lot of algorithms and data structures for similarity-aware searching with this kind of aims and applicable to our architecture. That is to say, The idea is to allow searching items based on any combination of the previous strategies, following a sublinear algorithm order. For example, classical alternatives include Burkhard-Keller Trees [58], Approximating and Eliminating Search Algorithm [59], Spatial Approximation Trees [60], Vantage Point Trees [61], Excluded Middle Vantage Point Forest [62], among others.

4.4 Search Strategy

The organizational memory contains a big data repository containing cases under the mentioned organization. A sequential searching in this kind of structure is not an option given the huge volume of involved data. However, a strategy based on the case structure is defined in order to smartly search for similar records. The strategy receives as input the symptoms loaded by the doctor in a narrative way, and from there the following stages are executed:

1. **Looking for Diagnosed Cases**: In this stage, only the cases with associated diagnoses are retrieved. Those cases without diagnostic are kept out because they do not present associated treatment which can be used for recommending.

Basically, this kind of query is performed under a MapReduce schema on a distributed cluster. Even, a good strategy could be to keep materialized views for optimizing the answer times.

2. **Stratified Sampling**: A stratified sampling based on the size of the previous answer is carried out in order to warranty the participation of all the level of risks (i.e. risk profile). Resampling techniques could be used in this stage for computing confidence intervals. The underlying idea is to get a representative profile of each kind of risk related to diseases contained in the organizational memory.

3. **Computing the Text Similarity**: In each obtained sample, the unstructured text incorporated in the query is contrasted to each case's symptom contained in the sample. Only when the text's similarity is upper or equal than a certain threshold (this is defined by the user), the case is retained, otherwise, it is discarded.

4. **Building the Vector Model**: Those sample's cases which had jumped the indicated threshold in the previous stage are expressed as a Boolean vector based on the ICD10 codes. Each dimension in the Boolean vector represents an ICD10 code, for that reason, it is expected a high dimensionality associated with each one of them.

5. **Computing the Distance Function**: In this stage, the pseudo-metric distance is calculated on the Boolean vector for each case. In other words, there are documents with similarity in symptoms representing each kind of risk profile and containing diagnostic. Based on the computed distance, a centroid is chosen to minimize the distance between points for each kind of risk profile, obtaining 5 centroids which are characteristic of each one. Only it will be retained those cases located within a given ratio considering the group's centroid as the center.

6. **Scoring**: For each risk profile, the remaining cases will be ordering in a descendent way by the additive. expression shown in Eq. 2:

$$score(a, b) = w_{ts} * ts(a[symptoms], b[symptoms]) + w_{md} * d(a_v, b_v) \qquad (2)$$

In Where:

"a" and "b" represents two cases from the organizational memory with the mentioned structure.

"$a[symptoms]$" and "$b[symptoms]$" represents a projection of the symptoms field in the record "a" and "b" respectively.

"$ts(a[symptoms], b[symptoms])$" represents the textual similarity computed on the symptoms field for the cases "a" and "b".

"a_v" and "b_v" represents the Boolean vector related to the cases "a" and "b".

"w_{ts}" indicates the weighting assigned to the text's similarity. This value must be contained in [0; 1].

"w_{md}" indicates the weighting assigned to the metric-distance function. This value must be contained in [0; 1], while the sum of w_{ts} and w_{md} must be equal to 1.

Those cases located on the first positions of the scoring result are associated with situations in which the treatments (i.e. previous experiences) could be considered like potential recommendations to be given in the answer. In addition, a synthesis using a frequency's histogram based on available treatments from the scoring list, grouped by risk profile, could be alternatively a complementary answer useful for recommendations.

5 Architecture: An Overview

Before entering in detail, it is important to highlight that the architecture is fed from medical data sources in which the doctors introduce the information on the health records under the form of diagnosis, symptoms, observations, etc. In that kind of information, the content (with the exception of the coded diagnostic) is expressed in a narrative way keeping in mind the plot informed from the patient. This implies that many aspects related to the colloquial language and informal use of it could be present (such as abbreviations, regional expressions, etc.), which incorporates an additional challenge at the moment in which two unstructured texts need to be analyzed in order to compute their similarity [63].

The architecture is fed from heterogeneous data sources which contain the electronic medical records, in where the record owner is the data source while the architecture just is limited to be a reader. So, the architecture is oriented to support text and diagnosis analysis but without be able to modify the original records. Basically, the architecture works with formatted and translated copies from records, avoiding overhead the primary system responsible for them.

The contents of the different data sources are consumed through the "*Data Collecting*" module. This module just read content from registered data sources, because of data sources need to be previously incorporated in order to define the data structure associated with the origin. This is reached through the "*Data Source Manager*" component as you can see in Fig. 1. Once the data source has declared the structure, the "*data pumpers*" component pump the data from the original data source, translating from the original data structure to the expected organizational memory structure introduced in Sect. 4.2. In addition, the "*Collecting Statistical Information*" component with collects information about the nature of each data source and their associated data in order to quantify typical problems related to the data quality (e.g. missing symptoms, missing diagnosis, etc.). The "*Data Sources Inspector*" component is the last responsible for indicating whether a given record will be incorporated in the organizational memory or not. In other words, if the translated record contains a patient ID, timestamp related to the medical visit, and explained symptoms, then they are kept in the organizational memory, otherwise it is discarded. It is important to highlight in this sense, that the missing content is tolerated in the diagnosis and treatments fields, because of it is possible to be in front of a new pathology.

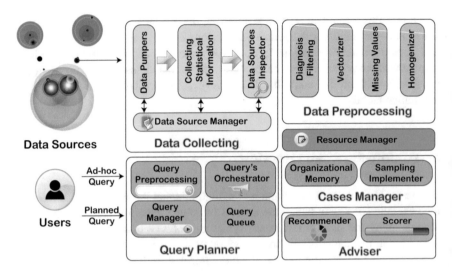

Fig. 1 A recommender system architecture

The "*Data Preprocessing*" module is associated with common functionalities required from the data collecting to the query planner. Basically, the "*Diagnosis Filtering*" component allows discriminating among the different kinds of incorporated diagnoses as support to the Text-similarity analysis. The "*Vectorizer*" component allows converting from an organizational memory record to a Boolean vector representation, which is useful for computing metric distances. The "*Missing Values*" component is able to identify different records with this kind of issue, proposing some alternatives for fixing it (e.g. the incorporation of values by analogy). The "*Homogenizer*" component is responsible for supporting the translation from the data structure registered in the data source manager to the expected data structure in the organizational memory.

The "*Cases Manager*" module is responsible for managing the organizational memory itself with the introduced data structure, and it incorporates a component which is responsible the sampling strategies, mainly thinking in the huge data volume related to it. Sampling and resampling techniques are especially useful when we want to optimize the data access time and obtain statistically valid results at the same time [64].

The "*Query Planner*" module is organized around four components, as follows (i) *Query Preprocessing*: it is responsible for capturing the ad hoc queries, organizing the query plan and put it on the query queue; (ii) *Query Manager*: it receives and registers the planned query, and put it into the query queue for execution in a periodic way; (iii) *Query Queue*: it keeps the ad hoc and planned queries in memory waiting for the moment in which each one has enough resources to be executed; (iv) *Query Orchestrator*: It negotiates with the "Resource Manager" module the needed resources for each query, and when they are obtained, it is the responsible for its execution and the return of results.

The "Adviser" module contains two components, as follows (i) *Scorer*: it is responsible for computing the score based on Eq. 2 on each case contained in the query result; (ii) "Recommender": Using the result from the scoring, it obtains the frequency of each contained treatment, providing them ordered in a descendant way by their frequency. Thus, the most frequent and related treatments are provided as an answer to similar cases.

The "Resource Manager" module is responsible for managing the available resources in order to receive new data, provide the proper storage, answer queries, and to provide recommendations.

With the aim of detailing internal aspects related to the architecture expected behavior, here each process by mean of the Business Process Model and Notation (BPMN) is described. Thus, this section is break-down into (i) Data Collecting Processes, (ii) Query Planner Processes, and (iii) Recommending and Scoring Process.

5.1 Data Collecting Processes

In general terms, the data collecting processes are constituted by two processes: (a) Data Sources Registering process which is responsible for incorporating new predefined data sources into a central register for future readings, and (b) The Process Data Pumping and Preprocessing associated with data gathering from data sources and the translation from each predefined data structure to the organizational memory structure.

Figure 2 describes the data sources registering process at a high level of granularity. It starts with searching for the new data sources which wish to be incorporated. If the

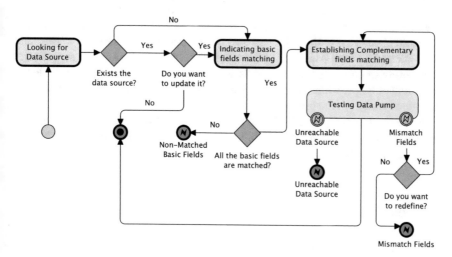

Fig. 2 The data sources registering process

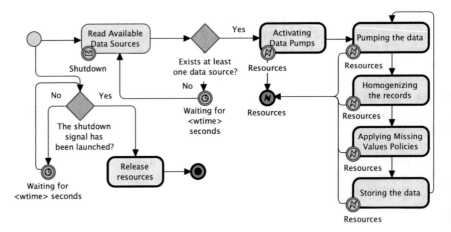

Fig. 3 The process of data pumping and preprocessing

data source previously exists, then the user is queried about whether to update it or not. When the data source is new or the user wants to update it, the basic fields need to be indicated (i.e. ID, timestamp and symptoms) jointly with the matching of fields between the data source and the target (i.e. the organizational memory). When the basic fields and its matching are verified, it is possible to establish the matching for complementary fields (i.e. narrative diagnostic, coded diagnostic, treatment, and risk profile). Once all the matching was defined, the data pump will verify the connection which could produce three different results: (i) The connection is successful and the data source is finally registered, (ii) The data source is unreachable and it is discarded, and (iii) There is no compatibility between the defined fields matching, which will imply a redefinition in the best of cases, or alternatively that the cases are discarded.

Figure 3 starts reading the defined data sources from the architecture. When there are no available data sources, the process keeps waiting for a time. The waiting time in the architecture is defined through the "*wtime*" parameter in seconds. When some data source definition is reached, immediately the associated data pumps are activated for starting to consume data. The data will be homogenized using the established matching of fields, while the missing values policies are applied, after that, the data are stored restarting the pumping of data.

From the data pumps' activation to the data storing is possible to experiment error messages associated with the availability of resources, which is especially critical in big data environments.

5.2 *Query Planner Processes*

The Query Planner component could be characterized through three processes: (i) Query registration, (ii) Query execution, and (iii) Ad hoc query performing. The query registration is responsible for verifying the planned queries and keeping them on the system. As you can see in Fig. 4, It starts with the query proposal which is verified in relation to the available resources and contents. Once the query has been verified, the execution requirements (e.g. volume of data to be retrieved, etc.) are established. Because there are previous queries defined, it is necessary to indicate the level of priority related to the new query and the rest of the queries. Thus, all the queries (i.e. the new query and the previous ones) are stored and updated in terms of the level of priority, which is especially critical at the moment in which the queries need to be launched.

Figure 5 describes the query execution process based on the defined queries on the architecture. Thus, it starts obtaining the planned queries from the catalog, and the launching of queries is initiated accordingly with the established priority. Thus,

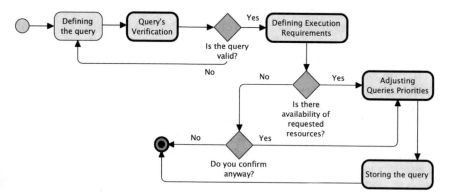

Fig. 4 The query registering process

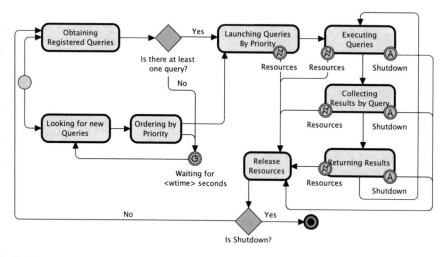

Fig. 5 The query execution process

the execution of queries, following the predefined priority, collecting and returning results in that order. In case of some kind of problem with the available resources, a signal is launched, releasing all the resources and restarting the querying schema again.

The second active thread in Fig. 5, continuously is looking for new queries. When some new queries are found, then they are ordered by priority and then they are launched following the same processing schema that others planned queries.

In this way, the process is able to run the planned queries following the established priorities, but also incorporating new planned queries reordering the queries under execution in terms of the newly established priorities. That is to say, a new query could change the priority order previously established, for example, because the last query is considered very important in terms of a given health criterium.

Finally, the process always will try to relaunch the queries even when the available resources are not completely enough. This is important because the new priorities can reorder the execution plan. That is to say, when a critical query is recently incorporated at the end of the list and the resources are not enough, the execution plan will be interrupted and relaunched guided by the new priorities.

The Ad hoc queries need to be incorporated, verified, launched but when the results are obtained, all the resources are released, and the query is not kept in the catalog (i.e. it is discarded).

Figure 6 synthesizes the ad hoc query performing process, which starts with the entering of the query. The query is verified following the criteria related to the available resources because the planned queries have priority on the ad hoc queries. In case of enough resources, the ad hoc query is launched and ran, the results are collected, and they are returned to the user. Once the results were returned, the engaged resources are released, while the query is discarded. This allows that the users enter the ad hoc queries that they want one and again, and when they want to reach some specific result in a reiterative way, they are able to introduce the query as a planned query.

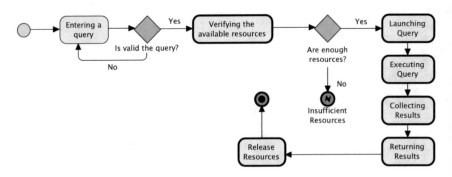

Fig. 6 The process of ad hoc query performing

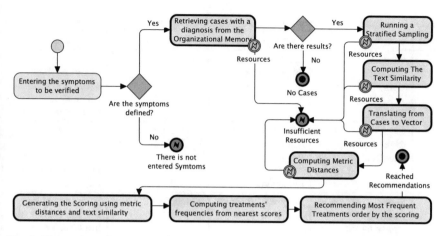

Fig. 7 The scoring and recommending process

5.3 The Scoring and Recommending Process

The "Scoring and Recommending" process is responsible for reaching the most common treatments for given symptoms incorporated in a narrative way by a doctor. Thus, the idea is that the doctor sends a text with a given diagnostic to the architecture, and it is able to obtain similar situations and based on them to recommend the associated treatments.

Figure 7 describes the mentioned process which starts through the entering of the diagnosis in a textual way. From there, a set of cases with diagnosis are obtained from the organizational memory. Next, a stratified sampling based on the risk profile is performed for warrantying the representativity in term of cases and level of risk. Once the sampling has been obtained, the text-similarity is contrasted using the entered symptom. From there, each case from the sampling is converted to a Boolean Vector model for computing the metric distances. Finally, the scoring is computed using Eq. 2, which allows weighting the text-similarity and metric distances.

Once the scoring is available, the architecture will identify the available treatments in the results, and then it will calculate the associated frequency of each one. In this way, the recommending will consist of the most frequent treatments associated with the most similar cases given the entered symptoms.

6 An Application Case

With the aim of providing an idea about the global strategy related to the architecture, this application case is synthesized for focusing on synthetical clinical stories and the need of looking for similar situations given an input. Table 2 provides data

Table 2 Simplified example of a database of clinical stories related to a level of risk 5

ID	Symptoms	Narrative diagnostic	ICD10 codes	Treatment
1	Weakness, Exhaustion, Loss of vigor, Permanent Tiredness	Possible chronic fatigue syndrome	G93.3	There is no treatment. Avoid stressing situations and mitigate symptoms
2	Chest Pain, Throat, Right Arm and Shoulders	Possible coronary heart disease (angina pectoris or myocardial infarction)	I20.0, I20.9 I21.1, I21.2, I21.3	If there is no relieving with nitroglycerin, then urgent attention is required
3	Sensation of Aire Absence or Drowning (Smoker Patient)	Dyspnea. Wheezing	R06.0, R06.2	Give up smoking. Request the full heart studies
4	Quick Weight losing not related to a given diet	Possible Crohn disease, or possible malignant neoplasm of an unspecified digestive organ	K50.0, K50.1, C78.8	Confirm differential diagnostic and derive to a specialist
5	Chest Pain and Burning. Stomach Burning. Pain in epigastrium	Possible angina pectoris. Possible gastroesophageal reflux disease	I20.0, I20.9, K21.0, K21.1	Confirm differential diagnostic and derive to a specialist
6	Loss of appetite for a long time	Possible anorexia. Possible anorexia nervosa	R63.0, F50.0	Studies for detecting gastric illness. In a negative case, apply psychotherapeutic and nutritional actions
7	Chest and Left Arm Pain. Fever. Heart Murmur	Possible myocardial infarction. Possible acute or subacute infective endocarditis	I33.0, I33.9, I21.1, I21.2, I21.3	Confirm differential diagnostic. Try with antibiotics
8	Appetite Absence, Queasiness, Vomit and Abdominal Pain	Possible acute gastritis	K29.0, K29.1	Confirm with gastroscopy. Personalized Diet
9	Chest, Left Hand, and Shoulders Pain. Burning in Sternum Zone	Possible acute myocardial infarction	I21.0, I21.1, I21.2, I21.3	Apply nitroglycerin with aerosol. If there are no positive results, immediately hospitalize
10	Sad and Dejected Patient, Loss of interest by its own daily habits	Possible depressive state	F32.0, F32.1	Prescribe antidepressant and start psychotherapy

Table 3 A synthetical clinical story defined as a query input

ID	Symptoms	Narrative diagnostic	ICD10 codes	Treatment
1	Chest Pain. Burning behind the sternum. Sudden pain on the left arm	Angina pectoris but possible myocardial infarction	I20.0, I20.9, I21.0, I21.1, I21.2, I21.3	Preventive Hospitalization

which contains real codes under the ICD10 with the purpose of being useful for exemplification, but they do not correspond to veridic patient information.

Now, a clinical story of a given patient is given in Table 3 simulating the input from a doctor who wants to get similar situations.

A searching phase from the organizational memory is started using a text-similarity analysis based on the unstructured text associated with the symptoms, narrative diagnostics, and treatments. Those records in the organizational memory without narrative diagnostics or treatments will be discarded. In this context, the searching retrieves the 2, 5, 7, and 9 records which are the most relevant in terms of the query.

The retrieved records are now represented as a vector using the ICD10 codes, considering each code as a dimension. Thus, when the code is present, that dimension will contain a number 1, otherwise will be a 0. Essentially, the vectors tend to be a high-dimensional vector (i.e. as many dimensions as ICD10 codes there exist), but really it is necessary to represent the present codes using a relationship (code, index). For example, "cs" represent the clinical story given as input (See Table 3), while r_i represents the corresponding record from Table 2:

- cs $= (I20.0, I20.9, I21.0, I21.1, I21.2, I21.3)$
- $r_2 = (I20.0, I20.9, I21.1, I21.2, I21.3)$
- $r_5 = (I20.0, I20.9, K21.0, K21.1)$
- $r_7 = (I21.1, I21.2, I21.3, I33.0, I33.9)$
- $r_9 = (I21.0, I21.1, I21.2, I21.3)$.

The retrieved vectors from the organizational memory are used for creating a structure oriented to search in a metric space. Some of that kind of structures are thought to operate based on discrete distance calculation (e.g. Burkhard-Keller Tree -BKT-), while other alternatives admit continuous distance (e.g. Vantage Point Tree - VPT-). In this example, the distance function to measure the distance between ICD10 vectors is the Euclidean distance, while the calculated values are continuous. A structure based on VPT could be used. The created structure serves as an index to search the most similar clinical story (or even the k-nearest neighbor).

In this example, a wide range searching would return the four indicated records (i.e. 2, 5, 7, and 9) as it is indicated below:

$$d(cs, r_2) = \|cs - r_2\| \tag{3}$$

$$d(cs, r_2) = \sqrt{\begin{array}{l}(1 - 1)^2_{I20.0} + (1 - 1)^2_{I20.9} + (1 - 0)^2_{I21.0} \\ +(1 - 1)^2_{I21.1} + (1 - 1)^2_{I21.2} + (1 - 1)^2_{I21.3}\end{array}} \tag{4}$$

$$d(sc, r_2) = 1 \tag{5}$$

Following the calculus in the same way that is expressed in Eqs. 3, 4, and 5 for the rest of the records, it is possible to obtain the following distances such as it is expressed in Eqs. 6, 7, and 8:

$$d(sc, r_9) = \sqrt{2} = 1.41 \tag{6}$$

$$d(sc, r_7) = \sqrt{5} = 2.24 \tag{7}$$

$$d(sc, r_5) = \sqrt{6} = 2.45 \tag{8}$$

As you can explicitly to appreciate in Eq. 4 in the distance calculus, the number of sub-radical is really the number of ICD10 that does not share both compared vectors (i.e. the number of ICD10 codes which are present in the first vector or in the second vector but not in both). Thus, a way to simplify the distance calculus is to limit to that strategy. For example, the Euclidean distance between sc and r_2 is $d(sc, r_2) = \sqrt{2}$, but simply it is possible to consider that $d(sc, r_2) = 2$. This variation allows in a natural way to use discrete structures in metric spaces for discrete distances.

In this way, and as you can see from the Eqs. 5, 6, 7, and 8 the lowest value is associated with the Eq. 5, which imply that is the most similar case in relation to the given input. For recommending purpose, the frequency of the treatments is computed from the result. Thus, in this case, "Nitroglycerin", "Confirm differential diagnostic" and "Urgent Attention" are derived such as the recommending associated with the results.

7 Conclusions

An architecture for e-Health Recommender Systems based on Similarity of Patients' Symptoms was introduced. This architecture registers the heterogeneous data sources, establishing the fields' matching in order to translate each case, from external repositories to a consolidated organizational memory. The organizational memory constitutes a big data repository fed through data pumps, which gather the original data, carry out the data pre-processing on data and incorporate them in a persistent storage.

The organizational memory allows establishing a common structure which allows analyzing the data in a consistent way, keeping in consideration the previous experiences and knowledge, which constitute a key asset for risk medical prevention in health systems related to smart cities.

In our architecture, the query could be outlined in different ways. On the one hand, it could be defined such as an unstructured text representing symptoms. On the other hand, it could be an electronic health record. Both cases pretend reach similar cases from the organizational memory, the first one tries to compare the probable diagnoses when it is unknow, while the second one is oriented to find a group of similar cases for contrasting diagnoses and symptoms. Both kinds of queries strategies are complementary, and they could be used in a concurrent way.

The underlying concepts and techniques related to calculating distance and text similarity were outlined. In addition, the processes' point of view related to the architecture was described through BPMN. An application case was presented with the aim of schematizing the global strategy associated with the architecture.

The architecture gathers in an articulated way, the integration among medical heterogeneous data sources, the capitalizing of previous experiences and knowledge through an organizational memory, the possibility to reach similar cases from unstructured text (i.e. symptoms) or an electronic health record, and the providing of recommendations based on the frequency of treatments and diagnosis coming from similar cases.

As future work, alternatives to the query strategy in order to improve the precision and accuracy will be analyzed.

References

1. Lee CH, Yoon H-J (2017) Medical big data: promise and challenges. Kidney Res Clin Pract 36(1):3–11
2. Yin Y, Zeng Y, Chen X, Fan Y (2016) The internet of things in healthcare: an overview. J Ind Inf Integr 1:3–13
3. Mezghani E, Exposito E, Drira K, Da Silveira M, Pruski C (2015) A semantic big data platform for integrating heterogeneous wearable data in healthcare. J Med Syst 39(12):185
4. Ivanović M, Budimac Z (2014) An overview of ontologies and data resources in medical domains. Expert Syst Appl 41(11):5158–5166
5. Degoulet P, Luna D, de Quiros FGB (2017) Chapter 6—Clinical information systems. In: Marin H, Massad E, Gutierrez M, Rodrigues R, Sigulem D (eds) Global health informatics. Academic Press-Elsevier, London, UK, pp 129–151
6. Watts NT (1989) Clinical decision analysis. Phys Ther 69(7):569–576
7. Ricci F, Rokach L, Shapira B (eds) (2015) Recommender systems handbook, 2nd edn. Springer, Boston, US
8. Abujar S, Hasan M, Hossain SA (2019) Sentence similarity estimation for text summarization using deep learning. In Kulkarni A, Satapathy S, Kang T, Kashan A (eds) ICDECT 2017: proceedings of the 2nd international conference on data engineering and communication technology. Advances in intelligent systems and computing, vol 828. Springer Singapore
9. Patil H, Thakur RS (2018) Document clustering: a summarized survey. In: I. Management Association (ed) Information retrieval and management: concepts, methodologies, tools, and applications. IGI Global, Hershey, PA, pp 47–64
10. Barón MJS (2017) Applying social analysis for construction of organizational memory of R&D centers from lessons learned. In: Proceedings of the 9th international conference on Information Management and Engineering, Barcelona, Spain, 9–11 Oct 2017, pp 217–220

11. Lee K, Kim Y, Joshi K (2017) Organizational memory and new product development performance: investigating the role of organizational ambidexterity. Technol Forecast Soc Change 120:117–129

12. Lamy JB, Sekar B, Guezennec G, Bouaud J, Séroussi B (2019) Explainable artificial intelligence for breast cancer: a visual case-based reasoning approach. Artif Intell Med 94:42–53

13. Munir K, Sheraz Anjum M (2018) The use of ontologies for effective knowledge modelling and information retrieval. Appl Comput Inf 14(2):116–126

14. Dessì D, Reforgiato Recupero D, Fenu G, Consoli S (2019) A recommender system of medical reports leveraging cognitive computing and frame semantics. In: Tsihrintzis G, Sotiropoulos D, Jain L (eds) Intelligent systems reference library, vol 149. Springer, Cham, pp 7–30

15. Kenter T, de Rijke M (2015) Short text similarity with word embeddings. In: Proceedings of the 24th ACM international on conference on Information and Knowledge Management—CIKM'15, Melbourne, Australia, 18–23 Oct 2015, pp 1411–1420

16. Kashyap A, Han L, Yus R, Sleeman J, Satyapanich T, Gandhi S, Finin T (2016) Robust semantic text similarity using LSA, machine learning, and linguistic resources. Lang Resour Eval 50(1):125–161

17. Mrabet Y, Kilicoglu H, Demner-Fushman D (2017) "TextFlow: a text similarity measure based on continuous sequences. In: Proceedings of the 55th annual meeting of the association for computational linguistics (volume 1: Long Papers), Vancouver, Canada, July 2017, pp 763–772

18. Yao L, Pan Z, Ning H (2019) Unlabeled short text similarity with LSTM encoder. IEEE Access 7:3430–3437

19. Miotto R, Weng C (2015) Case-based reasoning using electronic health records efficiently identifies eligible patients for clinical trials. J Am Med Inf Assoc 22(e1):e141–e150

20. Gómez-Vallejo HJ, Uriel-Latorre B, Sande-Meijide M, Villamarín-Bello B, Pavón R, Fdez-Riverola F, Glez-Peña D (2016) A case-based reasoning system for aiding detection and classification of nosocomial infections. Decis Support Syst 84:104–116

21. Su C-J, Huang S-F, Li Y (2018) Case based reasoning driven ontological intelligent health projection system. In: Proceedings of the 2nd international conference on Medical and Health Informatics—ICMHI'18, Tsukuba, Japan, June 2018, pp 185–194

22. Silva BN, Diyan M, Han K (2019) Big data analytics. Deep learning: convergence to big data analytics. SpringerBriefs in computer science. Springer, Singapore, pp 13–30

23. Navarro G (2014) Spaces, trees, and colors. ACM Comput Surv 46(4):1–47

24. Frittelli V, Diván MJ (2018) Clasificación de Modelos para Recuperación de Información. In: 6to. Congreso Nacional de Ingeniería Informática/Sistemas de Información (CoNaIISI 2018), Mar del Plata, Argentina, Nov 2018

25. Dominich S (2000) A unified mathematical definition of classical information retrieval. J Am Soc Inf Sci Technol 51(7):614–624

26. Baeza-Yates R, Ribeiro-Neto B (2011) Modern information retrieval: the concepts and technology behind search. Choice Rev Online 48(12):6950

27. Tenopir C (2008) Online systems for information access and retrieval. Libr Trends 56(4):816–829

28. Salton G, Fox EA, Wu H (1983) Extended boolean information retrieval. Commun ACM 26(11):1022–1036

29. Singh P, Dhawan S, Agarwal S, Thakur N (2015) Implementation of an efficient Fuzzy Logic based information retrieval system. ICST Trans Scalable Inf Syst 2(5):e5

30. Wong SK, Ziarko W, Wong PCN (1985) Generalized vector space model in information retrieval. In: Proceedings of the 8th annual international ACM SIGIR conference on Research and Development in Information Retrieval, Montreal, Quebec, Canada, 5–7 June 1985, pp 18–25

31. Furnas G, Deerwester S, Dumais S, Landauer T, Harshman R, Streeter L, Lochbaum K (1988) Information retrieval using a singular value decomposition model of latent semantic structure. In: Proceedings of the 11th annual international ACM SIGIR conference on Research and Development in Information Retrieval, Grenoble, France, 1988, pp 465–480

32. Robertson SE, Jones KS (1976) Relevance weighting of search terms. J Am Soc Inf Sci 27(3):129–146
33. Ribeiro BAN, Muntz R (1996) Belief network model for IR. In: SIGIR Forum (ACM Special Interest Group on Information Retrieval), pp 253–261
34. Turtle H, Croft WB (1989) Inference networks for document retrieval. In: Proceedings of the 13th annual international ACM SIGIR conference on Research and Development in Information Retrieval, SIGIR 1990, vol 51(2), pp 1–24
35. Robins D (2000) Interactive information retrieval: context and basic notions. Informing Sci 3(2):57–61
36. Agichtein E, Brill E, Dumais S (2006) Improving web search ranking by incorporating user behavior information. In: Proceedings of the twenty-ninth annual international ACM SIGIR conference on Research and Development in Information Retrieval, vol 2006, Seattle, Washington, USA, Aug 2006, pp 19–26
37. Frittelli V, Steffolani F, Teicher R, Picco J (2012) Búsqueda por similaridad aplicada en la recuperación de factores que inciden en el cálculo del índice de riesgo para la salud de la vivienda urbana. In: Rojas M, Meichtry N, Vázquez J (eds) Monitoreo de la salud ambiental análisis y perspectivas desde salud colectiva vulnerabilidad social y sistemas computacionales asociados. Instituto de Investigaciones Geohistricas de la Provincia del Chaco, Resistencia, pp 114–126
38. Losada DE, Barreiro A (2001) A logical model for information retrieval based on propositional logic and belief revision. Comput J 44(5):410–424
39. Lalmas M (1998) Logical models in information retrieval: introduction and overview. Inf Process Manag 34(1):19–33
40. Chen G, Zhang P (2012) The content extraction method of webpage information based on knowledge base. In: Proceedings of the 2012 5th international joint conference on Computational Sciences and Optimization, Harbin, Heilongjiang, China, 23–26 June 2012, pp 623–626
41. Zhang R, Ding G, Zhang F, Meng J (2017) The information retrieval technology of dynamic feedback artificial intelligence based on the neural network. In: Proceedings—2016 international conference on Smart City and Systems Engineering, Zhanggjiajie, Hunan, China, 25–26 Nov 2016, pp 250–253
42. Drias H, Khennak I, Boukhedra A (2009) A hybrid genetic algorithm for large scale information retrieval. In: Proceedings—2009 IEEE international conference on Intelligent Computing and Intelligent Systems, ICIS 2009 (1), pp 842–846
43. Sameer VU, Balabantaray RC (2014) Improving ranking of webpages using user behaviour, a genetic algorithm approach. In: 1st International conference on Networks and Soft Computing, Guntu, Andhra Pradesh, India, 19–20 Aug 2014—Proceedings, pp 1–4
44. Thakare AD, Dhote CA (2014) New unification matching scheme for efficient information retrieval using genetic algorithm. In: Proceedings of the 2014 international conference on Advances in Computing, Communications and Informatics, Delhi, India, 24–27 Sept 2014, pp 1936–1941
45. Bravo M, Montes A, Reyes A (2008) Natural language processing techniques for the extraction of semantic information in web services. In: 7th Mexican international conference on Artificial Intelligence—Proceedings of the Special Session, Atizapán de Zaragoza, México, 27–31 Oct 2008, pp 53–57
46. Calvillo EA, Mendoza R, Muñoz J, Martínez JC, Vargas M, Rodriguez LC (2016) Automatic algorithm to classify and locate research papers using natural language. IEEE Lat Am Trans 14(3):1367–1371
47. Ramli F, Noah SA, Kurniawan TB (2017) Ontology-based information retrieval for historical documents. In: 2016 3rd international conference on Information Retrieval and Knowledge Management, CAMP 2016—Conference Proceedings, Kuala Lumpur, Malaysia, 8–10 Aug 2016, pp 55–59
48. Jadhav PA, Chatur PN, Wagh KP (2016) Integrating performance of web search engine with machine learning approach. In: Proceeding of IEEE—2nd international conference

on Advances in Electrical, Electronics, Information, Communication and Bioinformatics, IEEE—AEEICB 2016, Chennai, Tamil Nadu, India, 27–28 Feb 2016, pp 519–524

49. Chávez E, Navarro G, Baeza-Yates R, Marroquín JL (2001) Searching in metric spaces. ACM Comput Surv 33(3):273–321

50. Amin S, Neumann G, Dunfield K, Vechkaeva A, Chapman KA, Wixted MK (2019) MLT-DFKI at CLEF eHealth 2019: multi-label classification of ICD-10 codes with BERT. In: Müller H, Cappellato L, Ferro N, Losada DE (ed) CEUR workshop proceedings, vol 2380. CEUR-WS

51. Muslim A, Mutiara AB, Suhendra A, Oswari T (2019) Expert mapping development system with disease searching sympthom based on ICD 10. In: 2018 Third international conference on informatics computing. IEEE, pp 1–4

52. Blanco A, Casillas A, Pérez A, Diaz de Ilarraza A (2019) Multi-label clinical document classification: impact of label-density. Expert Syst Appl 138:112835

53. Atutxa A, de Ilarraza AD, Gojenola K, Oronoz M, Perez-de-Viñaspre O (2019) Interpretable deep learning to map diagnostic texts to ICD-10 codes. Int J Med Inform 129:49–59

54. Idri A, Abnane I, Abran A (2015) Systematic mapping study of missing values techniques in software engineering data. In: 2015 IEEE/ACIS 16th international conference on Software Engineering, Artificial Intelligence, Networking and Parallel/Distributed Computing (SNPD), Takamatsu, Japan, 1–3 June 2015, pp 1–8

55. Kwak SK, Kim JH (2017) Statistical data preparation: management of missing values and outliers. Korean J Anesthesiol 70(4):407

56. Baeza-Yates R, Cunto R, Manber U, Wu S (1994) Proximity matching using fixed-queries trees. In: Lecture Notes in Computer Science (including subseries Lecture Notes in Artificial Intelligence and Lecture Notes in Bioinformatics), vol 807, LNCS. Springer Verlag, pp 198–212

57. Bozkaya T, Ozsoyoglu M (1997) Distance-based indexing for high-dimensional metric spaces. ACM SIGMOD Rec 26(2):357–368

58. Burkhard WA, Keller RM (1973) Some approaches to best-match file searching. Commun ACM 16(4):230–236

59. Micó ML, Oncina J, Vidal E (1994) A new version of the nearest-neighbour approximating and eliminating search algorithm (AESA) with linear preprocessing time and memory requirements. Pattern Recogn Lett 15(1):9–17

60. Navarro G (2002) Searching in metric spaces by spatial approximation. VLDB J 11(1):28–46

61. Yianilos P (1993) Data structures and algorithms for nearest neighbor search in general metric spaces. In: Proceedings of the fourth annual ACM-SIAM symposium on discrete algorithms, ACM, Austin, Texas, 25–27 Jan 1993, pp 311–321

62. Yianilos P (1999) Excluded middle vantage point forests for nearest neighbor search. In: DIMACS implementation challenge, ALENEX'99

63. Gupta A, Harrod M, Quinn M, Manojlovich M, Fowler K, Singh H, Saint S, Chopra V (2018) Mind the overlap: how system problems contribute to cognitive failure and diagnostic errors. Diagnosis 5(3):151–156

64. Bluhmki T, Dobler D, Beyersmann J, Pauly M (2019) The wild bootstrap for multivariate Nelson-Aalen estimators. Lifetime Data Anal 25(1):97–127

Printed in the United States
By Bookmasters